Lecture Notes in Mathematics

Edited by A. Dold and B. Eckmann

772

Jerome P. Levine

Algebraic Structure of Knot Modules

Springer-Verlag
Berlin Heidelberg New York 1980

Author

Jerome P. Levine
Department of Mathematics
Brandeis University
Waltham, MA 02154
USA

AMS Subject Classifications (1980): 13C05, 57Q45

ISBN 3-540-09739-2 Springer-Verlag Berlin Heidelberg New York
ISBN 0-387-09739-2 Springer-Verlag New York Heidelberg Berlin

Library of Congress Cataloging in Publication Data
Levine, Jerome P 1937-
Algebraic structure of knot modules.
(Lecture notes in mathematics; 772)
Bibliography: p.
Includes index.
1. Knot theory. 2. Modules (Algebra) 3. Invariants. I. Title. II. Series: Lecture notes in mathematics (Berlin); 772.
QA3.L28 no. 772 [QA612.2] 510s [514'.224] 80-246
ISBN 0-387-09739-2

Printing and binding: Beltz Offsetdruck, Hemsbach/Bergstr.
2141/3140-543210

TABLE OF CONTENTS

INTRODUCTION

In the study of n-dimensional knots, i.e. imbedded n-spheres in (n + 2)-space, one encounters a collection of Λ-modules $A_1, \ldots, A_n$ (the Alexander modules), where $\Lambda = Z[t, t^{-1}]$, the ring of integral Laurent polynomials. These modules encompass many of the classical knot invariants.

The more important properties and relations among these modules are more easily stated in terms of the Z-torsion submodules $\{T_i\}$ and the quotients $F_i = A_i/T_i$. An important additional feature is the existence of a product structure on F_q, when $n = 2q + 1$, and T_q, when $n = 2q$. It is now understood exactly which collections $\{T_i, F_i\}$ of Λ-modules, with product structure on the correct term, arise from knots (except for T_1). See [L] for more detail.

In the present work we make an algebraic study of the types of modules and product structures which arise as Alexander modules. In particular, we introduce a collection of invariants which are reasonably tractable but sensitive enough to reflect the panorama of these modules. In some cases, they succeed in classifying but we will be most concerned with determining when a given set of invariants can be realized.

A preliminary reduction of the problem is obtained as follows. Let π be an irreducible element of Λ. For any

Λ-module A we can then consider the π-primary submodule A_π. If A is a Z-torsion module, then we consider π which are integer primes; in this case A splits as the direct sum of the $\{A_\pi\}$. If A is Z-torsion free, we consider π which are irreducible primitive polynomials—but now A only <u>contains</u> the direct sum of the $\{A_\pi\}$. We will, in either case, concentrate on these π-primary modules. A further restriction will be made in the Z-torsion free case. The quotient ring $R = \Lambda/(\pi)$ is, in an obvious manner, a subring of the algebraic number field generated by a root of π. We will restrict our attention to the case when R is integrally closed, i.e. a Dedekind ring. Later on we will determine effective criterion for π to satisfy this condition.

The general setting then is the following. We consider a unique factorization domain Λ with a particular prime π such that the quotient ring $R = \Lambda/(\pi)$ is a Dedekind domain. If A is a π-primary Λ-module, we will derive from A a collection of R-modules $\{A_i, A^i\}$ tied together by means of a family of short exact sequences $0 \to A_{i+1} \to A_i \to A^i \to A^{i+1} \to 0$. It will also be useful to consider $\Delta_i = \mathrm{Cok}\{A_{i+1} \to A_i\} \approx \mathrm{Ker}\{A^i \to A^{i+1}\}$. If R is a Dedekind domain, these modules are all described by "numerical" invariants (rank and ideal class). When A carries a suitable product structure, there is a duality relationship between the $\{A_i\}$ and $\{A^i\}$ and, furthermore, Δ_i (or a closely related $\tilde{\Delta}_i$) inherits a more familiar type of product structure which can be handled by techniques from algebraic number theory.

We now outline in somewhat more detail the implementation of the above program. Our treatment of the Z-torsion case is relatively brief. In this case, R is the principal ideal domain $Z/p[t, t^{-1}]$ and all the derived modules are R-torsion. It is easy to see that the derived modules and sequences fail to classify A (except in trivial situations), but, on the other hand, the realizability problem is easily solved: All possible $\{A_i, A^i\}$, related by the required exact sequences, are realizable. When A has a product structure of the type we are considering, Δ_i inherits a symmetric (or skew-symmetric) bilinear form, as a vector space over Z/p, in which t acts isometrically. Such "isometric structures" are completely understood (see [La], [MI]). The product realizability theorem requires somewhat more work. It turns out that any $\{A_i\}$, with each $A_{i+1} \subseteq A_i$, together with any isometric structure on the $\Delta_i = A_i/A_{i+1}$ can be realized by some A with product structure. As mentioned above, because of duality relations between $\{A_i\}$ and $\{A^i\}$, this is the best one can hope for.

Z-torsion free $Z[t, t^{-1}]$-modules are treated as a special case of π-primary Λ-modules, where Λ is a unique factorization domain, $R = \Lambda/(\pi)$ is Dedekind, and the module has "π-only torsion," i.e. its annihilator is the principal ideal generated by some power of π. This corresponds precisely to demanding that the $\{A_i\}$ (or, in fact, just A_0) is R-torsion free. The realization theorem then states that any

$\{A_i, A^i\}$, where A_0 is R-torsion free, can be realized. The proof is long. As a first step, we consider the simplest case in which $A_0 = A_1 = \ldots = A_{d-1}$ and $A_d = 0$, for some d. These turn out (when A_0 is R-torsion free) to be exactly the projective $\Lambda/(\pi^d)$-modules. Realization of these modules reduces to the construction of <u>invertible</u> ideals in $S = \Lambda/(\pi^d)$, with a given reduction in $R = S/\pi S$. Once these elementary modules are realized, the general case is treated by amalgamating elementary modules together according to instructions read from the sequences: $0 \to A_{i+1} \to A_i \to A^i \to A^{i+1} \to 0$.

The ability of the derived modules and sequences to classify π-primary Λ-modules depends on the <u>degree</u>—the degree of A is the smallest d such that $\pi^{d+1}A = 0$. For modules of degree ≤ 3, classification is successful, but it is shown, by an example, that nonisomorphic modules of degree 4 can have isomorphic derived modules and sequences.

The product structures we consider are (skew)-Hermitian bilinear forms with values in Q/Λ, where Q is the quotient field, or, equivalently, in $S = \Lambda/\pi^{d+1}$, where d is the degree of the module. Such a structure will induce a (skew)-Hermitian form on $\tilde{\Delta}_i$ = R-torsion free quotient of Δ_i, with values in R. In the case $\Lambda = Z[t, t^{-1}]$, we are thus dealing with integral (skew)-Hermitian forms over algebraic number fields (see [J]). The classification question is handled by the following result. Two π-primary Λ-modules (satisfying an extra technical condition which is always true for knot modules) are isometric if and only if they are isomorphic in such a way that the induced isomorphisms on $\tilde{\Delta}_i$ are isometries. Thus the classification result above for degree ≤ 3 extends immediately to a classification result for modules with product structure.

To deal with the product realizability question, we restrict our attention to those modules which are the direct sum of "homogeneous" modules. A module of degree d is homogeneous if every nonzero element α arises from one of degree d in the sense that, for some $\lambda \in \Lambda$ relatively prime to π, we can write $\lambda\alpha = \pi^s\beta$, for some β with $\pi^d\beta \neq 0$. If $\Lambda = Z[t, t^{-1}]$, A is homogeneous of degree d if $A \otimes_Z Q$ is a free $Q[t, t^{-1}]/(\pi^{d+1})$-module—this condition can be easily expressed in terms of the Alexander polynomials. All modules of degree ≤ 2 are semi-homogeneous (direct sums of homogeneous modules), but modules of degree 3 are not necessarily. If A is homogeneous of degree d, we have $\{A_i\}$ all of the same rank—they can therefore be usefully considered to be "lattices" in the vector space $V = A_0 \otimes_R F$ - F is the quotient field of R. Furthermore $\tilde{\Delta}_{d-1}$ is the only nonzero $\tilde{\Delta}_i$, and since $\tilde{\Delta}_{d-1} = A_{d-1}$ is a lattice in V, the induced (skew)-Hermitian form on $\tilde{\Delta}_{d-1}$ determines such a form on V.

It turns out that <u>nonsingularity</u> of the original form is equivalent to the condition that A_i is dual to A_{d-1-i} in V for each i. Thus we can consider our invariants to consist of a nonsingular (skew)-Hermitian form over F and a nest of integral lattices $A_{d-1} \subseteq \cdots \subseteq A_k$ where $d = 2k$ or $2k + 1$

and A_k is self-dual if d is odd. Our realizability theorem then states that any such nest of integral lattices in the space of a nonsingular (skew)-Hermitian form can arise from a homogeneous module of degree d with a nonsingular product structure.

To obtain a more comprehensive realization theorem for nonsemihomogeneous modules we consider the "rational" invariants. When $\Lambda = Z[t, t^{-1}]$ this means we pass to $\bar{A} = A \otimes_Z Q$, considered as a module over the principal ideal domain $Q[t, t^{-1}]$; in our more general context, we pass to $\bar{A} = A \otimes_\Lambda \Lambda_\pi$, where Λ_π is the discrete valuation ring obtained by localizing Λ at (π). The derived invariants of $\bar{A}$ are, obviously, also invariants of A. The trivial nature of Λ_π immediately tells us that these invariants classify $\bar{A}$. When there is a product structure, the results of [MI] can be interpreted to state that the derived invariants (with the forms on $\bar{\Delta}_i$) classify $\bar{A}$ isometrically; the derived forms are (skew)-Hermitian forms over the algebraic number field F, when $\Lambda = Z[t, t^{-1}]$, which are well understood (see [La]). Realizability of these invariants by Λ_π-module is easily established ([MI]), so the problem is to pass from $\bar{A}$ to A. It turns out that realization corresponds to the existence of a self-dual lattice in $\bigoplus_i \bar{\Delta}_{2i}$, and this condition can be expressed in terms of the classical invariants of forms over F. As a by-product of this, one sees easily that semihomogeneous modules are relatively sparse, since semihomogenity requires that each $\bar{\Delta}_{2i}$ contain a self-dual lattice.

The final sections of this work are concerned with the ring $R = \Lambda/(\pi)$. The first problem is to determine, from π, whether R is integrally closed. We are, in fact, able to find a completely effective procedure involving prime factorization of π over Z/p, for each p dividing the discriminant of π, to resolve this issue. Once we know that R is Dedekind, we have the problem of computing the ideal class group of R. This is not the same as computing the ideal class group of an algebraic number field, since, if π is not monic, R contains nonintegers. R does, however, come close to being of the form $\mathcal{O}[\frac{1}{m}]$, where $\mathcal{O}$ is the ring of algebraic integers in F and m an integer. In fact, when π satisfies a condition first considered in [C], $R = \mathcal{O}[\frac{1}{m}]$, where m is the product of the first and last coefficients of π. In this case, the ideal class group of R can be determined from that of $\mathcal{O}$. This computation is then actually carried out, for some quadratic π, using the tables in [B].

§1. The derived exact sequences

Let Λ be an integral domain, and $\Pi \in \Lambda$ a prime element, i.e., if $\Pi = \Pi_1 \Pi_2$, then either Π_1 or Π_2 is a unit of Λ. Let A be a Λ-module. Define $K_i = K_i(A)$ to be the submodule of all elements killed by Π^i, i.e., $K_i = \{\alpha \in A: \Pi^i\alpha = 0\}$. Define $L_i \subseteq A$ to be $\Pi^i A$. We have inclusions:

$$0 = K_0 \subseteq K_1 \subseteq K_2 \subseteq \cdots$$

$$A = L_0 \supseteq L_1 \supseteq L_2 \supseteq \cdots$$

Finally define $A_i = K_{i+1}/K_i$, the <u>i-th lower Π-derivative</u> of A, and $A^i = L_i/L_{i+1}$, the <u>i-th upper Π-derivative</u> of A, for $i \geq 0$. Since $\Pi K_{i+1} \subseteq K_i$ and $\Pi L_i \subseteq L_{i+1}$, we conclude that A_i and A^i are modules over $\Lambda/(\Pi)$. Furthermore, multiplication by Π induces homomorphisms $\Pi_i: A_{i+1} \to A_i$ and $\Pi^i: A^i \to A^{i+1}$. We can also construct a homomorphism $\Delta_i: A_i \to A^i$ by multiplication by Π^i, since $\Pi^i K_{i+1} \subseteq \Pi^i A = L^i$, while $\Pi^i K_i = 0$.

These constructions are clearly functorial. Given a map $A \to B$ of Λ-modules, there are obvious induced maps $A_i \to B_i$, $A^i \to B^i$ commuting with Π_i, Π^i and Δ_i.

<u>Proposition 1.1</u>: The sequence

$$0 \to A_{i+1} \xrightarrow{\Pi_i} A_i \xrightarrow{\Delta_i} A^i \xrightarrow{\Pi^i} A^{i+1} \to 0$$

is exact for $i \geq 0$.

The proof is straightforward and will be omitted. We refer to this exact sequence as the i-th Π-primary sequence of A.

Note that $K = \bigcup_i K_i$ is the Π-primary submodule of A. If Λ is Noetherian and K is finitely generated, the nested sequence of $\{K_i\}$ terminates after a finite number of steps. The criterion for termination of $\{L_i\}$ is as follows:

Proposition 1.2: Let Λ be a Noetherian domain, Π a prime element of Λ and A a finitely generated Λ-torsion module. The following three conditions are equivalent.

i) $\Pi^m A = \Pi^{m+1} A$, for sufficiently large integer m.

ii) $A = A_\Pi \oplus \Pi^m A$, for sufficiently large integer m, where A_Π is the Π-primary submodule of A.

iii) There is an element $\phi \in \Lambda$ coprime to Π (i.e., $(\Pi, \phi) = \Lambda$) and an integer m, such that $(\Pi^m \phi)A = 0$ (and, therefore, $A = A_\Pi \oplus A_\phi$).

Proof:

(i) ==> (iii): Choose m large enough so that $\Pi^m A = \Pi^{m+1} A$ and $\Pi^m A_\Pi = 0$. Then $\Pi^m A \cap A_\Pi = 0$. Given $\alpha \in A$ we may find $\beta \in A$ such that $\Pi^m \alpha = \Pi^{2m} \beta$; the decomposition $\alpha = (\alpha - \Pi^m \beta) + \Pi^m \beta$ establishes $A = \mathrm{Ker}\Pi^m + \Pi^m A$.

(ii) ==> (iii): Choose m again so that $\Pi^m A_\Pi = 0$. Let $\alpha_1, \ldots, \alpha_k$ generate $\Pi^m A$; then $\alpha_i = \Pi^m(\sum_j \lambda_{ij}\alpha_j)$ for some $\lambda_{ij} \in \Lambda$. Rewriting this as $\sum_j (\delta_{ij} - \Pi^m \lambda_{ij})\alpha_j = 0$, we conclude that $\phi\lambda_i = 0$, for $\phi = \text{determinant } (\delta_{ij} - \Pi^m \lambda_{ij})$--see proof of [L, Cor. (1.3)]. Clearly $(\phi, \Pi^m) = \Lambda$, which implies $(\phi, \Pi) = \Lambda$.

<u>(iii) ⟹ (i)</u>: Choose m so that $\Pi^m A = 0$. Consider $\Pi^m \alpha$ an arbitrary element of $\Pi^m A$. If we write:

$1 = \lambda\phi + \mu\Pi$, then $\alpha = \lambda\phi\alpha + \mu\Pi\alpha$ and so

$\Pi^m\alpha = \lambda\phi\Pi^m\alpha + \mu\Pi^{m+1}\alpha = \mu\Pi^{m+1}\alpha$, which completes the proof.

The following propositions are of interest because of the definition of a module of type K (see [L]).

<u>Proposition 1.3</u>: If A is finitely generated and Π-primary, and λ is any element of Λ, then the following statements are equivalent:

i) Multiplication by λ defines an automorphism of A.

ii) Multiplication by λ defines an automorphism of every A_i.

iii) Multiplication by λ defines an automorphism of every A^i.

<u>Proof</u>: This follows by repeated use of the five lemma and the observations above.

<u>Proposition 1.4</u>: If A is Π-primary and Λ Noetherian, the following statements are equivalent:

i) A is finitely generated.

ii) A_i is finitely generated, for every i, and some $A_k = 0$.

iii) A^i is finitely generated, for every i, and some $A^k = 0$.

The proof follows immediately from the above observations.

Corollary 1.5: If $\Lambda = Z[t, t^{-1}]$ and A is Π-primary, the following statements are equivalent:

i) A is of type K.

ii) A_i is of type K, for every i, and $A_k = 0$ for some k.

iii) A^i is of type K, for every i, and $A^k = 0$ for some k.

§2. Finite modules

From now on we assume $\Lambda = Z[t, t^{-1}]$. We turn first to the case of finite Λ-modules. As usual any such can be decomposed into the direct sum of its p-primary components, p running over scalar primes. Each of these p-primary components is a Λ-module and so it suffices to study finite p-primary Λ-modules.

If we apply the considerations of §1 for $\Pi = p$, we have the family of p-primary sequences

$$0 \to A_{i+1} \xrightarrow{\Pi_i} A_i \xrightarrow{\Delta_i} A^i \xrightarrow{\Pi^i} A^{i+1} \to 0 \tag{2.1}$$

where each A_i, A^i is a $\Lambda_p = \Lambda/(p) = Z_p[t, t^{-1}]$-module. Since Λ_p is a principal ideal domain, we may describe the modules A_i, A^i by polynomial invariants. The condition that A be a module of type K is equivalent, by Corollary 1.4, to the condition that $t = 1$ not be a root of any of these polynomial invariants.

It is easy to see that the p-primary sequences (2.1) are not generally sufficient to classify A. For example, define two Λ-module structures on Z/p^2 by

i) $t\alpha = 2\alpha$,

ii) $t\alpha = (p + 2)\alpha \quad (p \neq 2)$.

It is easy to check that the p-primary sequences (2.1) are isomorphic, but the modules themselves, are not.

It is of interest to compare the p-primary sequences (2.1) of A and $e^2(A) = \mathrm{Ext}^2_\Lambda(A, \Lambda)$ in light of the duality relation ([L, 3.4(i)]). For any Λ_p-module B, define $B^* = \mathrm{Hom}_{Z_p}(B, Z_p)$ with Λ_p-module structure induced from that on B, i.e., if $\phi \in B^*$, $\lambda \in \Lambda_p$, then $\lambda\phi = \phi \circ \lambda$ (perhaps one really should set $\lambda\phi = \phi \circ \bar{\lambda}$). Then it is not difficult to check that $B^* \approx B$, if B is a finitely generated Λ_p-torsion module. The interest of $*$ is that it defines a <u>contra-variant</u> functor.

<u>Proposition 2.2</u>: Let A be a p-primary Λ-module of type K. Then $e^2(A)_i \approx A^i$, $e^2(A)^i \approx A_i$ and the i-th p-primary sequence for $e^2(A)$ is the "dual" of that for

$$A: 0 \to (A^{i+1})^* \xrightarrow{(\Pi^i)^*} (A^i)^* \xrightarrow{(\Delta_i)^*} A_i^* \xrightarrow{(\Pi_i)^*} A_{i+1}^* \to 0.$$

<u>Proof</u>: A homomorphism $e^2(A)_i \to (A^i)^*$ is defined as follows. Let $\phi \in \mathrm{Hom}_Z(A, Q/Z) \approx e^2(A)$ ([L, 4.2]) satisfy $p^{i+1}\phi = 0$. Then $\phi(p^{i+1}A) = 0$ and so ϕ induces a homomorphism $p^iA/p^{i+1}A \to Z_p \subseteq Q/Z$. It is a straightforward exercise to check it is bijective.

Similarly we define an isomorphism $e^2(A)^i \to (A_i)^*$. If $\phi \in p^i\mathrm{Hom}_Z(A, Q/Z)$, then $p^i\alpha = 0$ implies $\phi(\alpha) = 0$--i.e., $\phi(K_i) = 0$ where $K_i = \mathrm{Ker}\ p^i$ and we can define a homomorphism $K_{i+1}/K_i \to Z_p \subseteq Q/Z$. If $\phi \in p^{i+1}\mathrm{Hom}_Z(A, Q/Z)$, then $\phi(K_{i+1}) = 0$ and we get a well defined homomorphism $e^2(A)^i \to (A_i)^*$. Again, it is straightforward to check this is an isomorphism.

Finally we must check that the maps in the sequence (2.1) associated to $e^2(A)$, coincide, under these isomorphisms, with the duals of the homomorphisms in the sequence (2.1) associated to A. But, since these are all induced by multiplication by powers of p, this follows immediately from the naturality of the isomorphisms.

If we combine Proposition 2.2 with [L, 314(i)], we get the following result:

Corollary 2.3: Let $\{T_q\}$ be the Z-torsion submodule of the Alexander modules of an n-knot, and $T_{q,i}$, T_q^i the lower and upper p-derivative $\Lambda/(p)$-modules of T_q. Then we have the duality relations:

$$T_{q,i} \approx T^i_{n-q}$$

for $1 \leq q \leq n - 1$, and $T_q = 0$ otherwise. (See [G] for a related duality between polynomial invariants.)

§3. Realization of finite modules

We now solve the realization problem for the associated sequence (2.1).

Theorem 3.1: Let $\{B_i, B^i\}$ be a finite sequence of finitely generated torsion Λ_p-modules and:

(3.2) $$0 \to B_{i+1} \to B_i \to B^i \to B^{i+1} \to 0$$

a family of exact sequences, for $i \geq 0$. Then there exists a finitely generated p-primary Λ-module A, with p-primary derivatives $\{A_i, A^i\}$ and isomorphisms $A_i \approx B_i$, $A^i \approx B^i$, for every i, such that the p-primary sequences (2.1) correspond with (3.2) under these isomorphisms.

If each B_i, B^i is of type K, then so is A (by Corollary 1.5).

Proof: Notice that B_i and B^i are finite dimensional vector spaces over Z/p. The sequences (3.2) can be split (over Z/p) to give $B_i \approx B_{i+1} \oplus C_i$, $B^i \approx B^{i+1} \oplus C_i$. The automorphisms h_i, h^i of B_i, B^i induced by $\underset{\smile}{t}$ leave $B_{i+1} \subseteq B_i$ and $C_i \subseteq B^i$ invariant, and the automorphisms induced on C_i from h_i and h^i coincide. Thus we have a decomposition

$$B_0 \approx \bigoplus_{i \geq 0} C_i \approx B^0$$

where $B_k \approx \bigoplus_{i \geq k} C_i$ is invariant under h_0, and $C(k) = \bigoplus_{i \leq k} C_i$ is invariant under h^0. The automorphisms h_0, h^0 determine homomorphisms $h_{ij}, h^{ij}\colon C_j \to C_i$ and we have $h_{ij} = 0$ if $i < j$, $h^{ij} = 0$ if $i > j$ and $h_{ii} = h^{ii}$. It is clear that the vector space $\{C_i\}$ together with the homomorphisms $\{h_{ij}, h^{ij}\}$ are equivalent to the sequences (3.2).

Suppose we set $A = \bigoplus_{i > 0} D_i$, where D_i is a free Z/p^{i+1}-module. Let $E_i = D_i/pd_i$. An automorphism ϕ of A corresponds to a collection of homomorphisms $\phi_{ij}\colon D_j \to D_i$; if $i \geq j$, then $\phi_{ij}(D_j) \subseteq p^{i-j}D_i$. Define $\phi'_{ij}\colon E_j \to E_i$ induced from ϕ_{ij}; then $\phi'_{ij} = 0$ if $i > j$. If $i \geq j$, let $\phi''_{ij}\colon E_j \to E_i$ be the map induced from $(\frac{1}{p^{i-j}})\phi_{ij}$. Note $\phi_{ii} = \phi''_{ii}$. Now $A_k = \bigoplus_{j\geq k} p^{j-k}D_j/p^{j-k+1}D_j \sim \bigoplus_{j\geq k} E_j$; the automorphism of A_k induced by ϕ corresponds to the collection $\psi_{ij}: E_j \to E_i$ for $i, j \geq k$ where

$$\psi_{ij} = \begin{cases} 0 & i < j \\ \phi''_{ij} & i \geq j \end{cases}.$$

$A^k = \bigoplus_{j\geq k} p^k D_j/p^{k+1}D_j \approx \bigoplus_{j\geq k} E_j$ and ϕ induces an automorphism of A^k corresponding to $\phi'_{ij}: E_j \to E_i$ for $i, j, \geq k$.

In order for A to have p-primary sequences isomorphic to (3.2), we choose D_i so that $E_i \approx B_i$ --i.e. rank $D_i = \dim B_i$ --and ϕ_{ij} so that $\phi'_{ij} = h^{ij}$ and $\phi''_{ij} = h_{ij}$ for $i \geq j$. Lifting h_{ij}, h^{ij} to ϕ_{ij} is easy since the D_i are free over Z/p^{i+1}.

It will follow automatically that ϕ is an automorphism, from the fact that ϕ/A_k, for every k, is an automorphism--using the 5-lemma.

§4. Δ_i of finite modules

A weaker, but somewhat more tractable invariant of A is:

$$\Delta_i(A) = A_i/pA_{i+1} \approx \text{Kernel } \{p: A^i \to A^{i+1}\}$$

Note that A is of type K if and only if $\{\Delta_i(A)\}$ is a finite sequence of modules of type K (see 1.4).

These Λ_p-modules are independent and we have:

Corollary 4.1: If $\{\Delta_i\}$ is any finite sequence of finitely generated torsion Λ_p-modules, then there exists a finitely generated p-primary Λ-module A such that $\Delta_i(A) \approx \Delta_i$.

This follows immediately from Theorem 3.1, by setting $B_i = B_{i+1} \oplus \Delta_i$, $B^i = B^{i+1} \oplus \Delta_i$.

Corollary 4.2: Let $\{T_q\}$ be the Z-torsion submodules of the Alexander modules of an n-knot, and $\Delta_{i,q} = \Delta_i(T_q)$. Then we have $\Delta_{i,q} \approx \overline{\Delta}_{i,n-q}$, for $1 \leq q \leq n - 1$.

Conversely, given any finite collection $\{\Delta_{i,q}: q \neq 1, n/2, n - 1\}$ of Λ_p-modules of type K, satisfying $\Delta_{i,q} \approx \overline{\Delta}_{i,n-q}$, there exists an n-knot with these associated modules (see [G]).

Proof: It follows from Proposition 2.2 that $\Delta_i(e^2(A)) \approx \Delta_i(A)^*$, for any p-primary Λ-module of type K. Now the first part of 4.2 follows immediately from [L: 3.4(i)].

The second part follows from Corollary 4.1 and [L:11.1].

Corollary 4.2 provides a set of polynomial (over Z_p) invariants of a knot (see [G]) and a simple realization theorem for these invariants (except in dimensions 1, n/2, and n - 1). In the following sections we will discuss dimension n/2.

§5. Product structure on finite modules

Suppose A is finite and has a Z-linear, conjugate self-adjoint pairing $[,]:A \times A \to Q/Z$ (see [L:4.4]). The p-primary decomposition of A is also an orthogonal splitting with respect to [,]. If [,] is

$\in$-symmetric ($\in = \pm 1$) or non-singular, so are the restrictions of $[,]$ to each p-primary component. So we may assume A, itself, is p-primary.

We define an associated ϵ-symmetric Z-linear, conjugate self-adjoint pairing: $[,]_i: \Delta_i(A) \times \Delta_i(A) \to Z_p$, which is non-singular for all i if and only if $[,]$ is non-singular. For this purpose, let us identify $\Delta_i(A)$ with $K_{i+1}/(pK_{i+2} + K_i)$--see §1. If $\alpha, \beta \in \Delta_i(A)$, choose representatives $\bar{\alpha}, \bar{\beta} \in K_{i+1}$ and set $[\alpha, \beta]_i = p^i[\bar{\alpha}, \bar{\beta}] \in Z_p \subseteq Q/Z$ where Z_p is identified with the fractions whose denominator is p. It is straightforward to check that $[,]_i$ is well defined; for example, if $\bar{\alpha} = p\gamma$, the $p^i[\bar{\alpha}, \bar{\beta}] = p^{i+1}[\gamma, \bar{\beta}] = [\gamma, p^{i+1}\bar{\beta}] = 0$. It is immediate that $[,]_i$ is $\in$-symmetric, Z-linear, and conjugate self-adjoint.

Suppose $[,]$ is non-singular and $[\alpha, \beta]_i = 0$ for all $\beta \in \Delta_i(A)$. Therefore $p^i[\bar{\alpha}, \bar{\beta}] = 0$ for all $\bar{\beta} \in K_{i+1}$. We can use this fact to define a Z-homomorphism $\psi: p^{i+1}A \to Q/Z$ by the formula $\psi(p^{i+1}\xi) = [p^i\bar{\zeta}, \xi]$. since Q/Z is Z-injective, ψ extends to a homomorphism $A \to Q/Z$. By the non-singularity of $[,]$, there exists $\eta \in A$ such that $[\eta, \cdot]$ coincides with this homomorphism. In other words $[\eta, p^{i+1}\xi] = [p^i\bar{\alpha}, \xi]$, for all $\xi \in A$. Again by the non-singularity of $[,]$, we have $p^{i+1}\eta - p^i\bar{\alpha} = 0$, i.e., $\bar{\alpha} - p\eta \in K_i$. This proves $\bar{\alpha} \in K_i + pK_{i+2}$, and so $[,]_i$ is non-singular.

Now suppose each $[,]_i$ is non-singular. If $[,]$ were singular, then $[\alpha, A] = 0$ for some non-zero $\alpha \in A$. Let $\alpha = p^k\alpha_1$, where k is maximal; then $\alpha_1 \notin pA$ and $p^k[\alpha_1, A] = 0$. Since A is p-primary, there exists a positive integer i such that $p^{i+1}\alpha_1 = 0$ but $p^i\alpha_1 \neq 0$--obviously $i \geq k$. Now let $p^i\alpha_1 = p^{i+m}\alpha_2$, where m is maximal.

Therefore $\alpha_2 \in K_{i+m+1}$ and represents a non-zero element β of $\Delta_{i+m}(A)$. In fact, if $\alpha_2 \in K_{i+m}+pK_{i+m+2}$, then $p^i\alpha_1 = p^{i+m}\alpha_2 \in p^{i+m+1}K_{i+m+2}$, contradicting the maximality of m. We can now show that $[,]_{i+m}$ is singular, since $[\beta, \Delta_{i+m}(A)]_{i+m} = p^{i+m}[\alpha_2, K_{i+m+1}] = [p^{i+m}\alpha_2, K_{i+m+1}] = [p^i\alpha_1, K_{i+m+1}] = p^{i-k}[p^k\alpha_1, K_{i+m+1}] = p^{i-k}[\alpha, K_{i+m+1}] = 0$.

We have proved:

<u>Proposition 5.1</u>: Let A be a finite p-primary Λ-module with a Z-linear, ϵ-symmetric, conjugate self-adjoint pairing $[,]: A \times A \to Q/Z$. Then each $\Delta_i(A)$ inherits a Z_p-linear, ϵ-symmetric, conjugate self-adjoint pairing $[,]_i: \Delta_i(A) \times \Delta_i(A) \to Z_p$. Moreover $[,]$ is non-singular if and only if all $[,]_i$ are non-singular.

In the case $\epsilon = -1$ and p odd, it follows automatically that $[\alpha, \alpha] = 0$, for all $\alpha \in A$, and $[\alpha, \alpha]_i = 0$ for all $\alpha \in \Delta_i(A)$. If $p = 2$, then it is easy to check that $[\alpha, \alpha]_i = 0$ for all $\alpha \in \Delta_i(A)$ if <u>$i > 0$</u>. But, if A is of type K, then $[\alpha, \alpha] = 0$ for all $\alpha \in A$ (see [L: 13,5]), which implies $[\alpha, \alpha]_i = 0$, for all $\alpha \in \Delta_i(A)$, <u>$i \geq 0$</u>.

If A is of type K, we also have $[\alpha, \alpha]_i = 0$, for all $\alpha \in \Delta_i(A)$, $i \geq 0$, when $p = 2$ and $\epsilon = +1$ (see [L, 13.5(ii)]).

The adjoint of $[,]$ provides an isomorphism $A \approx \mathrm{Hom}_Z(A, Q/Z)$ and, according to Proposition 2.2, the associated sequence of Λ_p-modules $0 \to A_{i+1} \to A_i \to A^i \to A^{i+1} \to 0$ are isomorphic to their duals $0 \to (A^{i+1})^* \to (A^i)^* \to A_i^* \to A_{i+1}^* \to 0$. If we consider the sequence $0 \to A_{i+1} \to A_i \to \Delta_i(A) \to 0$ and $0 \to \Delta_i(A) \to A^i \to A^{i+1} \to 0$, it then follows that the latter are just the duals of the former.

From these considerations, the appropriate realization theorem for our purposes is:

<u>Theorem 5.2</u>: Let $0 \to B_{i+1} \to B_i \to \Delta_i \to 0$ be a collection of exact sequences of finitely generated torsion Λ_p-modules, and, on each Δ_i, let $[,]_i$ be a nonsingular ϵ-symmetric, Z_p-linear, conjugate self-adjoint pairing: $\Delta_i \times \Delta_i \to Z_p$, such that, for $p = 2$, $[\alpha, \alpha]_i = 0$ for every $\alpha \in \Delta_i$, $i \geq 0$.
Then there exists a finite p-primary Λ-module A with $B_i \simeq A_i$, $\Delta_i \simeq \Delta_i(A)$, such that the exact sequences $0 \to A_{i+1} \to A_i \to \Delta_i(A) \to 0$ are isomorphic to $0 \to B_{i+1} \to B_i \to \Delta_i \to 0$, and a non-singular, ϵ-symmetric, Z-linear, conjugate self-adjoint pairing $[,]: A \times A \to Q/Z$, inducing $[,]_i$ on $\Delta_i(A)$.

<u>Remark</u>: One cannot, in general, <u>prescribe</u> A and expect to lift $[,]_i$ to $[,]$. For example, if $A = Z/p^2$ and t = multiplication by $p - 1$, then $\Delta_0 = 0$, but $\Delta_1 \approx Z_p$ with $t = -1$. Clearly, every Δ_i admits a symmetric pairing $[,]_i$. But A admits no such pairing $[,]$, if p is odd, for, if so, then:
$(p - 1)[\alpha, \alpha] = [(p - 1)\alpha, \alpha] = [t\alpha, \alpha] = [\alpha, t^{-1}\alpha] = [\alpha, -(p + 1)\alpha]$ $= -(p + 1)[\alpha, \alpha]$. So $2p[\alpha, \alpha] = 0$, which implies $[,]$ is singular.

Proof: We translate into a matrix problem. Let $A = \bigoplus_{i \geq 0} D_i$, where D_i is a free module over $Z_{p^{i+1}}$ with a prescribed basis. Suppose we impose the condition on $[,]$ that the D_i be mutually orthogonal; then $[,]|D_i$ will be represented by a matrix γ_i, with entries in $Z_{p^{i+1}}$, which is ϵ-symmetric. If the action of t is represented by a matrix (γ_{ij}), where γ_{ij} represents the projection of $t|D_j$ onto D_i (see §3), then the conjugate self-adjointness corresponds to the matrix equation $(\gamma_{ij})(\gamma_i\delta_{ij})(\gamma_{ij})^\tau = (\gamma_i\delta_{ij})$, i.e.,

$$\sum_k \gamma_{ik}\gamma_k\gamma^\tau_{jk} = \gamma_i\delta_{ij} \tag{5.3}$$

The entries of these matrices are in different cyclic groups of prime power order and one needs to check that the multiplications in (5.3) are well defined, using the divisibility of γ_{ij} by p^{i-j} if $i > j$.

The pairing $[,]_i$ is represented by an ϵ-symmetric matrix λ_i; it is induced by $[,]$ if and only if $p^i\gamma_i = \lambda_i$, where $p^iZ_{p^{i+1}}$ is identified with Z_p. The action of t on A_0 is represented--as in §3--by a matrix (λ_{ij}), where $(\lambda_{ij})_{i,j\geq k}$ represents t on A_k, λ_{ii} is the matrix representative of t on $\Delta_i(A)$, and $\lambda_{ij} = 0$ if $i < j$. Then $\lambda_{ij} \equiv p^{j-i}\gamma_{ij} \bmod p$, for all i, j. Conjugate self-adjointness of $[,]_i$ corresponds to the matrix equation:

$$\lambda_{ii}\lambda_i\lambda^\tau_{ii} = \lambda_i$$

According to the hypotheses of Theorem 5.2, we are given λ_i, λ_{ij} as above, with the diagonal of λ_i zero when $p = 2$. We must construct γ_i, γ_{ij}, as above, satisfying (5.3), $p^i\gamma_i = \lambda_i$

and $\lambda_{ij} \equiv p^{j-i}\gamma_{ij}$ mod p. To begin, choose any ϵ-symmetric γ_i such that $p^i\gamma_i = \gamma_i$ and choose any γ_{ij}, for $i > j$, satisfying $\lambda_{ij} \equiv p^{j-i}\gamma_{ij}$ mod p. It remains to construct γ_{ij} for $i \le j$--the only conditions remaining are (5.3) and $\lambda_{ii} \equiv \gamma_{ii}$ mod p.

As a first approximation, choose $\gamma_i = 0$ for $i < j$ and <u>any</u> $\gamma_{ii} \equiv \lambda_{ii}$ mod p. Then (5.3) is satisfied <u>mod p</u>. Suppose that we have m-th order approximations γ_{ij} so that $\gamma_{ii} \equiv \lambda_{ii}$ mod p and (5.3) is satisfied <u>mod p^m</u>. We would now like to replace γ_{ij} by $\gamma'_{ij} = \gamma_{ij} + p^m\sigma_{ij}$, when $i \le j$, to get an $(m + 1)^{-st}$ order approximation. The condition on $\{\sigma_{ij}\}$ is:

$$\sum_k \sigma_{ik}\gamma_k\gamma_{jk}^\tau + \sum_k \gamma_{ik}\gamma_k\sigma_{jk}^\tau = \frac{(\gamma_i\delta_{ij} - \sum_k \gamma_{ik}\gamma_k\gamma_{jk}^\tau)}{p^m} \mod p$$

Since $\gamma_{ij} \equiv 0$ mod p, if $i \neq j$, $\gamma_k \equiv \lambda_k$ mod p, $\gamma_{ii} \equiv \lambda_{ii}$ mod p, this equation can be rewritten

$$(5.4(i, j)) \qquad \sigma_{ij}\lambda_j\lambda_{jj}^\tau + \lambda_{ii}\lambda_i\sigma_{ji}^\tau = \rho_{ij}$$

where ρ_{ij} is a collection of matrices satisfying $\rho_{ij} = \rho_{ji}^\tau$.

Clearly (5.4(i, j)) is ϵ times the transpose of (5.4(j, i)), so let us assume $i \le j$. If $i < j$, then $\sigma_{ji} = 0$; since λ_j and λ_{jj} are non-singular, mod p, we may solve for σ_{ij}(mod p). If $i = j$, we have the equation:

$$(\sigma_{ii}\lambda_i\lambda_{ii}^\tau) + (\sigma_{ii}\lambda_i\lambda_{ii}^\tau) = \rho_{ii}$$

When p is odd, since ρ_{ii} is ϵ-symmetric, we can find some matrix τ_i such that $\tau_i + \epsilon\tau_i^\tau = \rho_{ii}$ mod p. Since λ_i and λ_{ii} are non-singular mod p, we may then solve $\sigma_{ii}\lambda_i\lambda_{ii}^{\tau} = \tau_i$ for σ_{ii}, mod p.

When $p = 2$, we need to show that the diagonal of ρ_{ii} is even. To achieve this, we will need to be a little more careful. First of all, let us choose γ_i to have 0 diagonal--this can be done because the hypotheses of the theorem tell us that λ_i has 0 diagonal. Second of all, let us assume, as part of the condition of being an m-th order approximation, that each $\sum_j \gamma_{ij}\gamma_j\gamma_{ij}^{\tau}$ has diagonal entries divisible by 2^{m+1}. This implies directly that ρ_{ii} has even diagonal and, therefore, we may solve for τ_i and, so, σ_{ii}. In fact, we may choose τ_i rather freely, and, in particular, we may choose $\tau_i = \sigma_{ii}\lambda_i\lambda_{ii}^{\tau}$ to have arbitrary diagonal. We will show that a proper choice of τ_i will assure that $\sum_j \gamma'_{ij}\gamma_j(\gamma'_{ij})$ has diagonal entries divisible by 2^{m+2}, which will complete the inductive step. To see this

$$\sum_j \gamma'_{ij}\gamma_j\ (\gamma'_{ij})^{\tau} = \sum_j (\gamma_{ij}\gamma_j\gamma_{ij}^{\tau} + 2^m(\sigma_{ij}\gamma_j\gamma_{ij}^{\tau} + \gamma_{ij}\gamma_j\sigma_{ij}^{\tau}) + 2^{2m}\sigma_{ij}\gamma_j\sigma_{ij}^{\tau})$$

Since γ_j has 0 diagonal, $\sigma_{ij}\gamma_j\sigma_{ij}^{\tau}$ has even diagonal and so $2^{2m}\sigma_{ij}\gamma_j\sigma_{ij}^{\tau}$ has diagonal divisible by 2^{m+2} for any $m \geq 1$. Now we have, modulo 2^{m+2}:

$$\begin{aligned} \operatorname{diag}\sum_j \gamma'_{ij}\gamma_j\ (\gamma'_{ij})^{\tau}) &\equiv \sum_j (\operatorname{diag}(\gamma_{ij}\gamma_j\gamma_{ij}^{\tau}) + 2^m \operatorname{diag}(\sigma_{ij}\gamma_j\gamma_{ij}^{\tau} + \gamma_{ij}\gamma_j\sigma_{ij}^{\tau})) \\ &= \sum_j (\operatorname{diag}(\gamma_{ij}\gamma_j\gamma_{ij}^{\tau}) + 2^{m+1} \operatorname{diag}(\sigma_{ij}\gamma_j\gamma_{ij}^{\tau})) \end{aligned}$$

Since $\sum_j \operatorname{diag}(\gamma_{ij}\gamma_j\gamma_{ij}^{\tau})$ is divisible by 2^{m+1}, we may choose σ_{ij}, $i \neq j$, in any way and only take care that $\operatorname{diag}(\sigma_{ii}\gamma_i\gamma_{ii}^{\tau})$ have an appropriate value, mod 2. But $\sigma_{ii}\gamma_i\gamma_{ii}^{\tau} \equiv \tau_i$ mod 2, and we have completed the inductive step.

Since A is p-primary and finite, a m^{-th} order approximation for large enough m, will be the desired solution and the proof of Theorem 5.2 is complete.

§6. Classification of derived product structure

As a consequence of Theorem 5.2, we will consider finitely generated torsion Λ_p-modules V with an ϵ-symmetric non-singular conjugate self-adjoint pairing $[,]: V \times V \to Z_p$. A complete classification of such $(V, [,])$ can be easily derived from [M-1].

V admits a, unique up to isomorphism, orthogonal splitting $V = \sum_\phi V_\phi$, where ϕ ranges over some irreducible polynomials in Λ_p. V_ϕ is defined as follows:

i) If ϕ is relatively prime to $\bar{\phi}$, then V_ϕ is the sum of the ϕ-primary and $\bar{\phi}$-primary components of V, and $[,]|V_\phi$ is uniquely determined (up to isomorphism).

ii) If ϕ is a unit multiple of $\bar{\phi}$, then V_ϕ is the ϕ-primary component of V.

It, therefore, suffices to consider the case where V is ϕ-primary and ϕ is a unit multiple of $\bar{\phi}$. In this case V admits a, unique up to isomorphism, orthogonal splittine $V = \sum_i V_i$, where V_i is free over $\Lambda_p/(\phi^i)$. So we may assume V is free over $\Lambda_p/(\phi^r)$, for some $r \geq 1$.

If $\phi \neq t + 1$ or $t - 1$, we may, after multiplication by a suitable unit, assume $\phi = \bar{\phi}$. Then one can define a vector space $W = V/\phi V$ over $\Lambda_p/(\phi)$ and a non-singular Hermitian form $\langle,\rangle: W \times W \to \Lambda_p/(\phi)$--see [M-1, 3.3]. The isomorphism class of $(V, [,])$

is determined by that of $(W, \langle,\rangle)$, given r, and, conversely, any such $(W, \langle,\rangle)$ lifts to some $(V, [,])$. But $\Lambda_p/(\phi)$ is a finite field and, therefore $\langle,\rangle$ is uniquely, up to isomorphism, determined by W (see [M-1]).

When $\phi = t + 1$ or $t - 1$, we get a less trivial result, at least if $p \neq 2$. Again one defines $W = V/\phi V$, but now $\langle,\rangle$ is a $(-1)^{r-1}\epsilon$-symmetric, non-singular pairing on W, which is only a vector space over Z_p; isomorphism classes of $(V, [,])$, for a given r, are in one-one correspondense with isomorphism classes of $(W, \langle,\rangle)$.

The classification of $(W, \langle,\rangle)$ is well known $(p \neq 2)$. If $\epsilon = (-1)^r$, $\langle,\rangle$ is skew-symmetric and so W must be <u>even-dimensional</u>; $\langle,\rangle$ is then uniquely determined, up to isomorphism. If $\epsilon = (-1)^{r+1}$, $\langle,\rangle$ is symmetric and W can be any dimension. On each W there are two different $\langle,\rangle$ up to isomorphism, distinguished by their determinant in $(Z_p)^*/(Z_p^*)^2$ --see e.g. [Hi].

Note that, if V is a module of type K, then $\phi = t - 1$ is impossible and, if $p = 2$, $\phi = t + 1$ is also impossible. Thus, the above discussion covers all cases. We can summarize our observations.

<u>Theorem 6.1</u>: Let V be a finitely generated torsion Λ_p-module of type K. Then V admits a non-singular, ϵ-symmetric, conjugate self-adjoint pairing $V \times V \to Z_p$ if and only if $V \approx \bar{V}$ and, for every r such that $\epsilon = (-1)^r$, V contains an <u>even</u> number of summands of order $(t + 1)^r$.

The isomorphism classes of such pairing on V is in one-one correspondence with sequences $\{u_r\}$, where each $u_r \in Z_p^*/(Z_p^*)^2$, and u_r is defined for every r such that V has a non-zero summand of order $(t+1)^r$ and $\epsilon = (-1)^{r+1}$.

<u>Corollary 6.2</u>: Let Δ_{iq} be as in Corollary 4.2; so if $n = 2q$, then $\Delta_{iq} \approx \bar{\Delta}_{iq}$. In addition, Δ_{iq} contains an <u>even</u> number of summands of order $(t+1)^r$, for every $r \equiv q + 1 \bmod 2$. Conversely, given $q > 1$ and any finite collection of Λ_p-modules $\{\Delta_i\}$ with $\Delta_i \cong \bar{\Delta}_i$ such that each Δ_i contains an <u>even</u> number of summands of order $(t+1)^r$ for every $r \equiv q + 1 \bmod 2$, then there exists a $2q$-knot, such that $\Delta_{iq} \approx \Delta_i$ for all i.

Notice that the possible Alexander modules A_q for a $2q$-knot are <u>different</u> for q odd and for q even. For example, it is impossible that $A_q = T_q = Z_p$ with $t = -1$, when q is even, but it does occur for any q odd (see [Z] for $q = 1$). On the other hand, it is impossible that $A_q \approx \Lambda_p/((t+1)^2)$, for q odd, but it does occur for any q even.

§7. <u>Rational invariants</u>

We now turn our attention to Z-torsion free Λ-modules of type K. One technique for obtaining invariants of such modules is by passing to the rational extension--if A is a Z-torsion free Λ-module, then $\hat{A} = A \otimes_Z Q$ is a $\Lambda \otimes_Z Q = \Gamma$ module of type K (i.e., finitely generated and $t - 1$ is an automorphism). Since $\Gamma = Q[t, t^{-1}]$ is a principal ideal domain, $\hat{A}$ is classified by its <u>invariant factors</u>.

If we write $\hat{A} \approx \bigoplus_{i=1}^{m} \Gamma/(\gamma_i)$ where γ_1 is not a unit and γ_i/γ_{i+1} then the (γ_i) are uniquely determined and we refer to them as the invariant factors. The following easy theorem completely determines these "rational invariants" of Z-torsion free modules of type K.

Theorem 7.1: If $\gamma_1,\ldots,\gamma_m$ are the invariant factors of a Z-torsion free module of type K, then γ_i can be chosen to be integral (i.e., $\gamma_i \in \Lambda \subseteq \Gamma$) satisfying $\gamma_i(1) = 1$. Conversely any collection of non-units $\{\gamma_1,\ldots,\gamma_m\} \subseteq \Lambda$ satisfying: γ_i/γ_{i+1} for $i = 1,\ldots,m-1$, $\gamma_i(1) = 1$, are the rational invariant factors of a Z-torsion free module of type K.

Proof: We may certainly choose γ_i to be primitive integral. It then suffices to prove $\gamma_m(1) = 1$. Consider the annihilator I of $A = \{\lambda \in \Lambda : \lambda A = 0\}$. This is a principal ideal for if λ is the greatest common divisor of the elements of I, then $m\lambda \in I$ for some non-zero integer m. But $m\lambda A = 0$ implies $\lambda A = 0$, since A is Z-torsion free. It is easy to see that λ will also generate the annihilator of $\hat{A}$--thus $(\lambda) = (\gamma_m)$. So it suffices to check $\lambda(1) = \pm 1$. In fact, it suffices to find any $\mu \in \Lambda$ such that $\mu(1) = \pm 1$ and $\mu A = 0$, since, of necessity, $\lambda | \mu$.

Let $\alpha_1,\ldots,\alpha_k$ be generators of A. Since $t-1$ is an automorphism, we may write $\alpha_i = (t-1)\Sigma\lambda_{ij}\alpha_j$ or:

$$\sum_{j=1}^{k} ((t-1)\lambda_{ij} - \delta_{ij})\alpha_j = 0; \quad i = 1,\ldots,k$$

But then $\mu = \det((t-1)\lambda_{ij} - \delta_{ij})$ will be the desired element.

(See [L, Cor. 1.3].)

Conversely, it is easy to check that $A = \bigoplus_{i=1}^{m} \Lambda/(\gamma_i)$ is a Z-torsion free module of type K.

§8. Z-torsion-free modules

If (λ) is the annihilator ideal of A, we refer to λ as a minimal polynomial of A. For each prime factor π of λ, we may consider the π-primary sequences of A:

$$0 \to A_{i+1} \to A_i \to A^i \to A^{i+1} \to 0$$

constructed in §1 where A_i, A^i are the lower and upper i-th π-derivatives of A. Recall that $A_i = 0$, for large enough i, and $A^i = 0$, for large enough i, if and only if all of the other prime factors of λ are coprime to π. This follows from Proposition 1.2. In this case A is the sum of its π-primary submodules, over all prime π, and the π-primary sequences are invariants entirely of the π-primary submodule. Therefore in our considerations it will suffice to consider π-primary modules, i.e., λ is a power of a single prime π. Another restriction we will make concerns the structure of the ring $\Lambda/(\pi)$, which we will denote by R. In order to extract invariants from the π-primary sequences and consider classification questions, we will want to assume R is a Dedekind domain.

Definition: If Λ is a domain, then an irreducible element is Dedekind if the domain $R = \Lambda/(\pi)$ is a Dedekind domain.

In §27ff. we will consider the problem of deciding which polynomials in $\Lambda = Z[t, t^{-1}]$ are Dedekind.

The question of whether a Z-torsion free module A is of type K will not concern us further because of:

Proposition 8.1: A finitely generated Z-torsion free module A is of type K if and only if its minimal polynomial λ satisfies the condition $\lambda(1) = \pm 1$.

Proof: The necessity of the condition follows from the proof of Theorem 7.1. On the other hand, if $\lambda(1) = 1$, we may write $\lambda = (t - 1)\mu \pm 1$. Since $\lambda A = 0$, this implies $t - 1$ is an automorphism.

Thus if we consider π-primary modules, they will be of type K if and only if $\pi(1) = \pm 1$.

§9. Π-only torsion

There is one restriction imposed upon the π-primary sequences of a Z-torsion free Λ-module of type K which we must take note of.

Proposition 9.1: If A is a π-primary Λ-module and A_i the i-th lower π-derivative of A, then A is Z-torsion free if and only if A_0 (and, therefore, every A_i) is Z-torsion free.

Since $A_0 = \mathrm{Ker}\{\pi: A \to A\}$ is a submodule of A, necessity is clear. Conversely, if α is a non-zero Z-torsion element of A, choose k so that $\pi^k\alpha \neq 0$ and $\pi^{k+1}\alpha = 0$. Then $\pi^k\alpha$ is a non-zero Z-torsion element of A_0.

<u>Proposition 9.2</u>: If π is a non-constant irreducible element of Λ and $R = \Lambda/(\pi)$, then an R-module is Z-torsion free if and only if it is R-torsion free.

<u>Proof</u>: Since $Z \subseteq R$ as a subring, R-torsion free implies Z-torsion free. Conversely, suppose α is an R-torsion element. Choose $\lambda \in \Lambda$, corresponding to a nonzero element of R which annihilates α. Since λ is not divisible by π, the ideal in Λ generated by (λ, π) contains a non-zero scalar m. Clearly $m\alpha = 0$.

Since it will be simpler (and more general) to work with a general integral domain Λ, rather than only $\Lambda = Z[t, t^{-1}]$, we correspondingly broaden the preceding notions. Given an irreducible element $\pi \in \Lambda$, let Λ_π be the localization at π, i.e., the subring of the quotient field $Q(\Lambda)$ with denominations <u>not</u> divisible by π. Then Λ_π is a discrete valuation ring with residue field $\Lambda_\pi/\pi\Lambda_\pi \approx Q(R)$. We say a π-primary Λ-module A has <u>π-only</u> torsion if $\lambda\alpha = 0$ implies $\alpha = 0$ or π/λ.

<u>Proposition 9.3</u>: The following are equivalent:

a) A has π-only torsion.

b) $A \to A \otimes_\Lambda \Lambda_\pi$ is injective.

c) A_0 is R-torsion free ($R = \Lambda/(\pi)$)

Corollary 9.4: If $\Lambda = Z[t, t^{-1}]$ and π is non-constant, then A is Z-torsion free if and only if A has π-only torsion.

We leave the proofs to the reader.

§10. Statement of realization theorem

Our realization theorem will imply that there are no further restrictions on the π-primary sequences, at least when π is Dedekind.

Theorem 10.1: Let π be a Dedekind element of the noetherian domain Λ, $R = \Lambda/(\pi)$ and $\{0 \to A_{i+1} \to A_i \to A^i \to A^{i+1} \to 0\}$ $(i = 0,\ldots,k)$ a family of exact sequences of finitely generated R-modules such that $A_{k+1} = 0 = A^{k+1}$ and A_0 is R-torsion free. Then there exists a π-primary finitely generated Λ-module A whose π-primary sequences are isomorphic to the given ones.

More precisely, if $0 \to A_{(i+1)} \to A_{(i)} \to A^{(i)} \to A^{(i+1)} \to 0$ are the π-primary sequences of A, we require isomorphisms $A_{(i)} \approx A_i$, $A^{(i)} \approx A^i$ so that the diagram:

$$\begin{array}{ccccccccc}
0 \to & A_{(i+1)} & \to & A_{(i)} & \to & A^{(i)} & \to & A^{(i+1)} & \to 0 \\
& \wr\wr & & \wr\wr & & \wr\wr & & \wr\wr & \\
0 \to & A_{i+1} & \to & A_i & \to & A^i & \to & A^{i+1} & \to 0
\end{array}$$

are commutative.

The proof will occupy the next several sections. The procedure for constructing A will be step-by-step, proceeding inductively on k.

We will assume a suitable candidate for Kernel π^k already exists and adjoin to it a certain particularly simple type of "building block" module to create the desired A. The implementation of this program falls into two parts:

i) Understanding the relationship between the π-primary sequences of A and $K_k(A) = \text{Ker}\ \pi^k$, where $\pi^{k+1}A = 0$.

ii) Constructing the "building blocks."

§11. Inductive construction of derived sequences

We first look at (i). Let A be a π-primary Λ-module such that $\pi^{k+1}A = 0$, and let $B = K_k(A)$. Let $\{0 \to A_{i+1} \to A_i \to A^i \to A^{i+1} \to 0\}$ and $\{0 \to B_{i+1} \to B_i \to B^i \to B^{i+1} \to 0\}$ be the π-primary sequences of A, B respectively. We first show how to directly extract those of B from those of A.

Define $C_i = A_i$ for $i < k$, $C_i = 0$ for $i \geq k$, and maps $C_{i+1} \to C_i$ (which are obviously injective) to correspond to those on A_{i+1}, when $i < k - 1$, and zero when $i \geq k - 1$. We now define C^i, maps $C^i \to A^i$ and $C_i \to C^i \to C^{i+1}$, by downward recursion on i, so that the diagram:

$$\begin{array}{ccccccccc} 0 & \to & C_{i+1} & \to & C_i & \to & C^i & \to & C^{i+1} & \to & 0 \\ & & \downarrow & & \downarrow & & \downarrow & & \downarrow & & \\ 0 & \to & A_{i+1} & \to & A_i & \to & A^i & \to & A^{i+1} & \to & 0 \end{array}$$

is commutative with exact rows.

Let $C^i = 0$ for $i \geq k$ and the maps $C^i \to A^i$, $C_i \to C^i \to C^{i+1}$, of course, be zero. Now assume C^i and the various maps are defined for $i > \ell$ where $\ell > k$. Consider the diagram:

$$\begin{array}{ccccccccc} 0 \to & C_{\ell+1} & \to & C_\ell & \dashrightarrow & C^\ell & \dashrightarrow & C^{\ell+1} & \to 0 \\ & \downarrow & & \downarrow & & \vdots & & \downarrow & \\ 0 \to & A_{\ell+1} & \to & A_\ell & \to & A^\ell & \to & A^{\ell+1} & \to 0 \end{array}$$

where C_ℓ and the dotted maps emanating from it are to be defined. If $\ell = k - 1$, then $C_{\ell+1} = C^{\ell+1} = 0$ and we define $C_\ell = C^\ell$. The map $C_\ell \to C^\ell$ is the identity and $C^\ell \to A^\ell$ is defined by $C^\ell = C_\ell = A_\ell \to A^\ell$. If $\ell < k - 1$, then define C^ℓ to be the <u>push-out</u> of:

$$\begin{array}{ccc} A^\ell & \to & A^{\ell+1} \\ & & \uparrow \\ & & C^{\ell+1} \end{array}$$

with the two dotted maps $C^\ell \dashrightarrow C^{\ell+1}$ and $C^\ell \dashrightarrow A^\ell$ also thus defined. By property of push-outs, the map $A_\ell \to A^\ell$ lifts to a unique map $A_\ell \to C^\ell$ whose composition with $C^\ell \to C^{\ell+1}$ is zero. Since $C_\ell = A_\ell$, this defines the map $C_\ell \to C^\ell$. It is routine to check that the resulting diagram is commutative with exact rows.

<u>Lemma 11.1</u>: The π-primary sequences of $B = K_k(A)$ are isomorphic to $\{0 \to C_{i+1} \to C_i \to C^i \to C^{i+1} \to 0\}$ constructed above, so that the maps $C_i \to A_i$, $C^i \to A^i$ correspond to those induced by the inclusion $B \to A$.

<u>Proof</u>: It is clear that $B_i = A_i$ for $i < k$, $B_i = 0$ for $i \geq k$,

so that we have isomorphisms $C_i \approx B_i$ with the desired properties. Furthermore $B^i = 0$ for $i \geq k$. Proceeding by downward induction on i, we observe that $B_{k-1} \approx B^{k-1}$ and if $\ell < k - 1$, then B^ℓ is isomorphic to the push-out of

$$\begin{array}{ccc} A^\ell & \to & A^{\ell+1} \\ & & \uparrow \\ & & B^{\ell+1} \end{array}$$

This follows from the commutative diagram

$$\begin{array}{ccccccccc} 0 & \to & B_{\ell+1} & \to & B_\ell & \to & B^\ell & \to & B^{\ell+1} & \to & 0 \\ & & \downarrow & & \downarrow & & \downarrow & & \downarrow & & \\ 0 & \to & A_{\ell+1} & \to & A_\ell & \to & A^\ell & \to & A^{\ell+1} & \to & 0 \end{array}$$

together with the fact that $B_\ell \to A_\ell$ and $B_{\ell+1} \to A_{\ell+1}$ are isomorphisms. The lemma now follows readily.

§12. Inductive recovery of derived sequences

We now turn around and try to describe the π-primary sequence of A from those of B. Of course we will need some additional information. We have seen that $B_i = A_i$ for $i < k$ and $B_i = 0$ for $i \geq k$. Thus A_k is lost when we pass from A to B; the additional information will correspond precisely to A_k and its relation to the π-primary sequences of B. There is a string of epimorphisms $B^0 \to B^1 \to \dots \to B^{k-1}$ and an injection $A_k \to A_{k-1} = B_{k-1} \approx B^{k-1}$--the latter an isomorphism because $B_k = 0 = B^k$. This injection $A_k \to B^{k-1}$ lifts to a map $A_k \to B^0$ via the composite epimorphism $B^0 \to B^{k-1}$. In fact, since

$A_k = \dfrac{\operatorname{Ker} \pi^{k+1}}{\operatorname{Ker} \pi^k} = \dfrac{A}{B}$, $B^0 = \dfrac{B}{\pi B}$, and $\pi A \subseteq B$, there is an obvious map $A_k \to B^0$ induced by multiplication by π. It follows directly from the definitions of the maps involved that this will serve as the desired lift.

Given the family of exact sequences $\{0 \to B_{i+1} \to B_i \to B^i \to B^{i+1} \to 0\}$ such that $B_k = B^k = 0$, and a submodule $C \subseteq B_{k-1} \approx B^{k-1}$ together with a map $\phi \colon C \to B^0$ which makes the following diagram commutative:

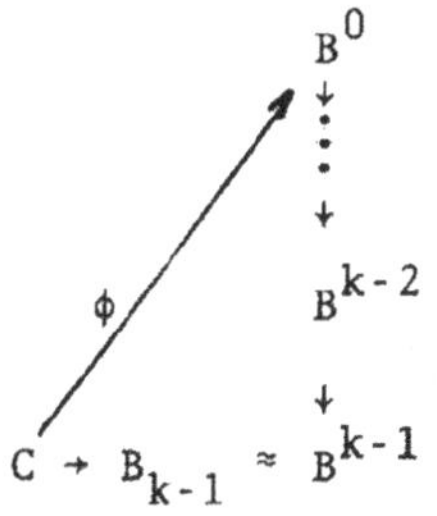

we construct a family of exact sequences $\{0 \to C_{i+1} \to C_i \to C^i \to C^{i+1} \to 0\}$ and commutative diagrams:

$$\begin{array}{ccccccccc} 0 & \to & B_{i+1} & \to & B_i & \to & B^i & \to & B^{i+1} & \to & 0 \\ & & \downarrow & & \downarrow & & \downarrow & & \downarrow & & \\ 0 & \to & C_{i+1} & \to & C_i & \to & C^i & \to & C^{i+1} & \to & 0 \end{array}$$

Let $C_i = B_i$ for $i < k$, $C_k = C$, $C_i = 0$ for $i > k$. The injections $C_{i+1} \to C_i$ are the induced ones from $\{B_i\}$, for $i < k - 1$, the given injection $C_k = C \to B_{k-1} = C_{k-1}$, for $i = k - 1$, and zero for $i \geq k$. Let $\phi \colon C \to B^i$ be the composition $C \overset{\phi}{\to} B^0 \twoheadrightarrow B^i$. Then define $C^i = C \oplus \operatorname{Cok} \phi_i$, for $i < k$, $C^k = C$ and $C^i = 0$ for $i > k$. The epimorphism $B^i \to B^{i+1}$ induces $\operatorname{Cok} \phi_i \to \operatorname{Cok} \phi_{i+1}$ and, therefore, epimorphisms $C^i \to C^{i+1}$ for $i < k-1$. For $i = k-1$, there is an obvious projection $C^{k-1} = C \oplus \text{Cok}\ \phi \to C = C^k$ and, for $i \geq k$, of course, this map must be zero. A map $B^i \to C^i$ is clearly defined by projection onto $\text{Cok}\ \phi_i$ when $i < k$, and zero when $i \geq k$.

To define $C_i \to C^i$ we proceed as follows. For $i < k$, $C_i = B_i$ and the projection $B_i \to \text{Cok}\ \phi_i \subseteq C^i$ will induce the desired map. For $i = k$, we have $C_k = C = C^k$ and we take the identity map. For $i > k$, we take zero.

With these definitions, commutativity of the diagram

$$\begin{array}{ccccccccc} 0 \to & B_{i+1} & \to & B_i & \to & B^i & \to & B^{i+1} & \to 0 \\ & \downarrow & & \downarrow & & \downarrow & & \downarrow & \\ 0 \to & C_{i+1} & \to & C_i & \to & C^i & \to & C^{i+1} & \to 0 \end{array}$$

and exactness of the bottom row is a straightforward exercise. We wish to prove that the sequences $\{0 \to C_{i+1} \to C_i \to C^i \to C^{i+1} \to 0\}$ are, in fact, the π-primary sequences of A when $\{0 \to B_{i+1} \to B_i \to B^i \to B^{i+1} \to 0\}$ are those of $B = \text{Ker}\ \pi^k$ and ϕ is constructed as above. This will follow from a more general fact--that the second construction is the inverse of the first, provided ϕ, C are chosen appropriately. More specifically, suppose $\{0 \to A_{i+1} \to A_i \to A^i \to A^{i+1} \to 0\}$ is a family of exact sequences, with $A_i = 0 = A^i$, for $i > k$. We then construct a new family of exact sequences $\{0 \to B_{i+1} \to B_i \to B^i \to B^{i+1} \to 0\}$ by the construction of §11. Define $C = A_k$ and $\phi_i\colon C \to B^i$, inductively as follows: $\phi_i = 0$ for $i \geq k$, of course. Let ϕ_{k-1} be the composite injection $C = A_k \to A_{k-1} = B_{k-1} \overset{\approx}{\to} B^{k-1}$. If ϕ_{i+1} has been defined, we lift it to ϕ_i satisfying the commutative diagram:

$$\begin{array}{ccc} & \overset{\phi_i}{\nearrow} & B^i \\ & & \downarrow \\ C & \xrightarrow{\phi_{i+1}} & B^{i+1} \end{array}$$

Recall that B^i is the push-out:

$$\begin{array}{ccc} B^i & \to & B^{i+1} \\ \downarrow & & \downarrow \\ A^i & \to & A^{i+1} \end{array}$$

Inductively, assume the composition $C \xrightarrow{\phi_{i+1}} B^{i+1} \to A^{i+1}$ is zero. Then, by properties of push-outs, there is a unique lift of ϕ_{i+1} to ϕ_i satisfying the extra inductive hypothesis. Now, let $\phi = \phi_0$.

<u>Proposition 12.1</u>: If A_k is projective, then the exact sequences $\{0 \to C_{i+1} \to C_i \to C^i \to C^{i+1} \to 0\}$ constructed from $\{0 \to B_{i+1} \to B_i \to B^i \to B^{i+1} \to 0\}$ and ϕ, according to the instructions of 12, are isomorphic to $\{0 \to A_{i+1} \to A_i \to A^i \to A^{i+1} \to 0\}$.

<u>Proof</u>: We first prove that ϕ_i, constructed above, fits into an exact sequence:

$$(*) \qquad 0 \to A_k \xrightarrow{\phi_i} B^i \to A^i \to A^k \to 0 \quad \text{for} \quad i < k$$

where $B^i \to A^i$ is produced in the construction of §11, and $A^i \to A^k$ is the composition of the epimorphisms $A^i \to A^{i+1} \to \ldots \to A^k$.

We proceed by downward induction on i. If $i = k - 1$, ϕ_{k-1} corresponds to the homomorphism $A_k \to A_{k-1}$ in such a way that the exact sequence $0 \to A_k \to A_{k-1} \to A^{k-1} \to A^k \to 0$ corresponds to the sequence (*), thereby proving it exact. Now consider the commutative diagram:

$$(**)\qquad \begin{array}{ccccccccc}
 & & & & 0 & & 0 & & \\
 & & & & \downarrow & & \downarrow & & \\
 & & & & B_{i+1} & \to & A_{i+1} & & \\
 & & & & \downarrow & & \downarrow & & \\
 & & & & B_i & \to & A_i & & \\
 & & & & \downarrow & & \downarrow & & \\
0 \to & A^k & \xrightarrow{\phi_i} & & B^i & \to & A^i & \to & A^k \to 0 \\
 & \| & & & \downarrow & & \downarrow & & \| \\
0 \to & A_k & \xrightarrow{\phi_{i+1}} & & B^{i+1} & \to & A^{i+1} & \to & A^k \to 0 \\
 & & & & \downarrow & & \downarrow & & \\
 & & & & 0 & & 0 & &
\end{array}$$

The columns and the bottom row are exact, and we want to prove the top row exact. Since $i < k - 1$, the maps $B_i \to A_i$ and $B_{i+1} \to A_{i+1}$ are isomorphic according to the construction of §11. It is now an exercise in diagram-chasing to prove that the exactness of the top row is implied by these properties of diagram (**)--<u>except</u> for the fact that the composition $A_k \xrightarrow{\phi_i} B^i \to A^i$ is zero. But this is the property of ϕ_i that was made part of our inductive assumptions during its construction.

Now $A_k \approx A^k$ is assumed projective and we may, therefore, split the exact sequence (*) and write:

$$A^i \approx A^k \oplus \mathrm{Cok}\,\phi_i = C^i$$

Furthermore, if we choose these splittings "consistently," i.e., choose a splitting for $i = 0$ and compose with the epimorphism $A^0 \to A^1 \to \cdots \to A^i$ to obtain the splitting of the other i, then it follows from (**) that the maps $A^i \to A^{i+1}$ agree with the constructed maps $C^i \to C^{i+1}$. Furthermore it is clear that the maps $B^i \to A^i$ correspond to projection on $\mathrm{Cok}\,\phi_i$, which is exactly the definition of $B^i \to C^i$. The remaining identifications of $\{0 \to A_{i+1} \to A_i \to A^i \to A^{i+1} \to 0\}$ with $\{0 \to C_{i+1} \to C_i \to C^i \to C^{i+1} \to 0\}$ are straightforward, since $A_i = B_i = C_i$ for $i < k$, and $C_k = C = A_k$.

<u>Corollary 12.2</u>: Let A be a finitely generated Λ-module with π-only torsion, $\pi \in \Lambda$ a Dedekind element such that $\pi^{k+1}A = 0$, and $B = \mathrm{Ker}\,\pi^k$. If $\{0 \to C_{i+1} \to C_i \to C^i \to C^{i+1} \to 0\}$ are constructed from the π-primary sequences of B, as described in §12, using $C = A_k = A/B$ and $\phi\colon A/B \to B^0 = B/\pi B$ induced by multiplication by π, then the π-primary sequences of A are isomorphic to $\{0 \to C_{i+1} \to C_i \to C^i \to C^{i+1} \to 0\}$.

<u>Proof</u>: This will follow from Proposition 12.1, if we identify our present ϕ with that in Proposition 12.1 (which we will now denote by ϕ'). We show, by downward induction on i, that $\phi_i\colon C \to B^i$ corresponds to the map $\phi_i'\colon A/B \to \pi^iB/\pi^{i+1}B$ induced by multiplication by π^{i+1}. For $i = k-1$, ϕ_{k-1} is defined as the composition $C = A_k \to A_{k-1} = B_{k-1} \to B^{k-1}$, but recall that the first of these maps

is induced by multiplication by π, while the second is induced by multiplication by π^{k-1}. For the inductive step, recall that ϕ_i is the unique lift of $\phi_{i+1} = \phi'_{i+1}$ whose composition with $B^i \to A^i$ is zero. But ϕ'_i is certainly a lift of ϕ'_{i+1} and the composition $A/B \xrightarrow{\phi'_i} \pi^i B/\pi^{i+1}B \to \pi^i A/\pi^{i+1}A$, which is again the map induced by multiplication by π^{i+1}, is certainly zero. This completes the proof.

§13. Homogeneous and elementary modules

We now construct the "building blocks."

We first define a notion of π-primary Λ-modules. Consider the module $A_\pi = A \otimes_\Lambda \Lambda_\pi$ over Λ_π (see §9). It is easy to see that A_π uniquely decomposes as a direct sum of modules isomorphic to $\Lambda_\pi/(\pi^i)$, for various values of i, if Λ is a noetherian domain, because then Λ_π is a local domain, all of whose ideals are of the form (π^i).

If $\Lambda = Z[t, t^{-1}]$, then the isomorphism class of A_π is equivalent to the isomorphism class of $A \otimes_Z Q$ (see §7).

Definition: We say a finitely generated π-primary Λ-module is homogeneous of degree d if and only if A_π is a free $\Lambda_\pi/(\pi^d)$ module.

Proposition 13.1: A is homogeneous if and only if every non-zero i-th lower π-derivative A is finitely generated and all have the same rank, as R-modules. The degree of A is the largest d such that $A_{d-1} \neq 0$.

Proof: Consider the lower π- derivatives of A_π. Since $\Lambda_\pi/(\pi) = Q(R)$, the quotient field of $R = \Lambda/\pi$, $(A_\pi)_i$ is a vector space over $Q(R)$. It is clear that A_π is a free $\Lambda_\pi/(\pi^d)$-module of dimension k if and only if $(A_\pi)_i$ are vector spaces of rank k for $0 \leq i < d$, and $(A_\pi)_i = 0$ for $i \geq d$. The proposition now follows from the observation that $(A_\pi)_i \approx A_i \otimes_R Q(R)$, which is a consequence of the exactness of the localization functor.

Thus a finitely generated π-primary A is homogeneous of degree d if and only if $A_i = 0$ for $i \geq d$ and the injections $\pi_i \colon A_{i+1} \to A_i$ have R-torsion cokernels when $i < d - 1$ or, equivalently, $A^i = 0$ for $i \geq d$ and the surjections $\pi^i \colon A^i \to A^{i+1}$ have R-torsion kernels for $i < d - 1$.

Definition: A finitely generated π-primary Λ-module A will be called elementary if it has π-only torsion and $\pi A = \text{Ker } \pi^{d-1}$, for some d.

Proposition 13.2: A finitely generated π-primary Λ-module A with π-only torsion is elementary if and only if every injection $\pi_i \colon A_{i+1} \to A_i$ is either zero or onto. Furthermore A is homogeneous of degree d, where d is the integer in the definition.

Proof: The cokernel of $\pi_i \colon A_{i+1} \to A_i$ is isomorphic to $\text{Ker } \pi^{i+1}/\pi\text{Ker } \pi^{i+2} + \text{Ker } \pi^i$. If $\pi A = \text{Ker } \pi^k$, then $\pi\text{Ker } \pi^{i+1} = \text{Ker } \pi^i$ for any $i \leq k$, and so $\pi_i \colon A_{i+1} \to A_i$ is onto for $i < k$. On the other hand $A_{k+1} = 0$ since $\pi^{k+1}A = 0$. Conversely if $A_i = 0$ for $i > k$, then $\pi^{k+1}A = 0$. If $\pi_i : A_{i+1} \to A_i$ is onto for $i < k$, then $\operatorname{Ker} \pi^{i+1} = \pi \operatorname{Ker} \pi^{i+2} + \operatorname{Ker} \pi^i$ for $i < k$. By induction on i, we see that $\operatorname{Ker} \pi^{i+1} \subseteq \pi \operatorname{Ker} \pi^{i+2}$ for $i < k$ and, therefore, $\operatorname{Ker} \pi^k \subseteq \pi \operatorname{Ker} \pi^{k+1} = \pi A$. The last statement of the proposition follows from Proposition 13.1.

So the π-primary sequences of an elementary module of homogeneous degree d reduce to isomorphisms $A_0 \approx A_1 \approx \dots \approx A_{d-1} \approx A^{d-1} \approx \dots \approx A^1 \approx A^0$.

§14. Realization of elementary modules

The elementary π-primary module will be our "building block."

Lemma 14.1: Let M be any finitely generated R-torsion free R-module, d any positive integer. Then if π is Dedekind, there exists an elementary π-primary module A of homogeneous degree d, with $A^0 \approx M$.

Proof: If M is free, the lemma is trivial, since we can choose for A the free S-module, where $S = \Lambda/(\pi^d)$, of the same rank as M. Furthermore, since $(A' \oplus A'')^0 \approx (A')^0 \oplus (A'')^0$, direct sums of elementary modules are elementary, and any $M \approx M' \oplus I$, where M' is free and I is an ideal of R, it suffices to prove:

Given any ideal I of R, there exists an ideal $\bar{I}$ of S, such that $\bar{I} \equiv I \bmod \pi S$ (we regard $R = S/\pi S$), and $\pi \bar{I} = \bar{I} \cap \pi S$.

Since I is generated by two elements, we may attempt to define $\bar{I}$ by choosing $\bar{\alpha}, \bar{\beta} \in S$ such that

i) $\bar{I} \equiv I \bmod \pi S$.

We will successively modify $\{\bar{\alpha}, \bar{\beta}\}$ so that after the k-th modification, $\bar{I}$ will also satisfy:

ii)$_k$ $\bar{I} \cap \pi S = \pi\bar{I} + \bar{I} \cap \pi^{k+1}S$.

We begin at $k = 0$, which is automatic, and the proof will be complete when $k = d$.

Assume, therefore, that $\bar{I}$ satisfies (i) and (ii)$_k$ for some $k \geq 0$. We will replace $\bar{\alpha}$ by $\bar{\alpha}' = \bar{\alpha} + \bar{\alpha}''$, and $\bar{\beta}$ by $\bar{\beta}' = \bar{\beta} + \bar{\beta}''$ for suitable $\bar{\alpha}'', \bar{\beta}'' \in \bar{I} \cap \pi^{k+1}S$. If $\bar{I}'$ denotes the ideal generated by $\{\bar{\alpha}', \bar{\beta}'\}$, then $\bar{I}'$ will satisfy (i). We will use the following notation. If λ, μ, x, y, etc. denote elements of R, then $\bar{\lambda}, \bar{\mu}, \bar{x}, \bar{y}$, etc. will denote some element of S (or Λ) in their residue classes mod πS (or $\pi\Lambda$). Conversely, if $\bar{\lambda}, \bar{\mu}, \bar{x}, \bar{y} \in S$ or Λ, then λ, μ, x, y denote their residue classes in R.

Let K be the set of all pairs (λ, μ) such that $\bar{\lambda}\bar{\alpha} + \bar{\mu}\bar{\beta} \in \pi S$ --this criterion depends only on λ, μ. We have an obvious short exact sequence:

$$0 \to K \to R \oplus R \to I \to 0$$

where $R \oplus R \to I$ is the map $(\lambda, \mu) \mapsto \bar{\lambda}\alpha + \bar{\mu}\beta \bmod \pi S$. Since π is Dedekind, the sequence splits and K is an R-torsion free R-module of rank one. The pair $(\beta, -\alpha)$ obviously belongs to K; thus there is another pair (λ_0, μ_0) which, together with $(\beta, -\alpha)$, generates K.

The four elements $\alpha, \beta, \lambda_0, \mu_0$, generate R, for if they generated a proper ideal J of R, then $K \subseteq J \oplus J$ --this is impossible, since K is a direct summand. Write:

$$1 = x\alpha + y\beta + \sigma\lambda_0 + \tau\mu_0$$

The element $\bar{\gamma} = \bar{\lambda}_0\bar{\alpha} + \bar{\mu}_0\bar{\beta} \in \pi S \cap \bar{I} = \pi\bar{I} + \bar{I} \cap \pi^{k+1}S$, by $(ii)_k$. Therefore we can choose $\bar{\theta} \in \bar{I} \cap \pi^{k+1}S$ so that $\bar{\gamma} - \bar{\theta} \in \pi\bar{I}$. Now set $\bar{\alpha}'' = -\bar{\sigma}\,\bar{\theta}$, $\bar{\beta}'' = -\bar{\tau}\,\bar{\theta}$. We will check $(ii)_{k+1}$ for $\bar{I}' = (\bar{\alpha}', \bar{\beta}')$. Suppose $\bar{\lambda}\,\bar{\alpha}' + \bar{\mu}\,\bar{\beta}' \in \bar{I}' \cap \pi S$. Since $\bar{\alpha}''$, $\bar{\beta}'' \in \pi S$, $\bar{\lambda}\,\bar{\alpha} + \bar{\mu}\,\bar{\beta} \in \pi S$, and so $(\lambda, \mu) \in K$. We may write $(\lambda, \mu) = \xi(\beta, -\alpha) + \eta(\lambda_0, \mu_0)$, for some $\xi, \eta \in R$.. Thus modulo $\pi\bar{I}$,

$$\begin{aligned}
\bar{\lambda}\,\bar{\alpha}' + \bar{\mu}\,\bar{\beta}' &= (\bar{\xi}\,\bar{\beta} + \bar{\eta}\,\bar{\lambda}_0)(\bar{\alpha} - \bar{\sigma}\,\bar{\theta}) + (-\bar{\xi}\,\bar{\alpha} + \bar{\eta}\,\bar{\mu}_0)(\bar{\beta} - \bar{\tau}\,\bar{\theta}) \\
&= \bar{\xi}(\bar{\beta}\,\bar{\alpha} - \bar{\beta}\,\bar{\sigma}\,\bar{\theta} - \bar{\alpha}\,\bar{\beta} + \bar{\alpha}\,\bar{\tau}\,\bar{\theta}) + \bar{\eta}(\bar{\lambda}_0\bar{\alpha} - \bar{\lambda}_0\bar{\sigma}\,\bar{\theta} + \bar{\mu}_0\bar{\beta} - \bar{\mu}_0\bar{\tau}\,\bar{\theta}) \\
&= \bar{\xi}(\bar{\alpha}\,\bar{\tau} - \bar{\beta}\,\bar{\sigma})\bar{\theta} + \bar{\eta}(\bar{\lambda}_0\bar{\alpha} - \bar{\mu}_0\bar{\beta} - \bar{\theta}(\bar{\lambda}_0\bar{\sigma} + \bar{\mu}_0\bar{\tau})) \\
&= \bar{\xi}(\bar{\tau}\,\bar{\alpha} - \bar{\sigma}\,\bar{\beta})\bar{\theta} + \bar{\eta}(\bar{\lambda}_0\bar{\alpha} + \bar{\mu}_0\bar{\beta} - \bar{\theta}(1 - \bar{x}\,\bar{\alpha} - \bar{y}\,\bar{\beta})) \\
&= \bar{\xi}(\bar{\tau}\,\bar{\alpha} - \bar{\sigma}\,\bar{\beta})\bar{\theta} + \bar{\eta}((\bar{\gamma} - \bar{\theta}) + \bar{\theta}(\bar{x}\,\bar{\alpha} + \bar{y}\,\bar{\beta})) \\
&= (\bar{\alpha}(\bar{\xi}\,\bar{\tau} + \bar{\eta}\,\bar{x}) + \bar{\beta}(-\bar{\xi}\,\bar{x} + \bar{\eta}\,\bar{y}))\bar{\theta}
\end{aligned}$$

But $\bar{\theta} \in \pi^{k+1}S$ and so $\bar{\theta}\,\bar{\alpha}$, $\bar{\theta}\,\bar{\beta} \in \pi^{k+1}\bar{I} \subseteq \pi\bar{I}$. Thus we have proved $\bar{I}' \cap \pi S \subseteq \pi\bar{I}$. But $\bar{\theta} \in \pi^{k+1}S$ which implies $\bar{I} = \bar{I}' \bmod \pi^{k+1}S$ and so $\pi\bar{I} \subseteq \pi\bar{I}' + \pi^{k+2}S$. Thus $(ii)_{k+1}$, and so Lemma 14.1, is proved.

§15. Classification of elementary modules

<u>Lemma 15.1</u>: Suppose A is an elementary π-primary Λ-module of degree d and A^0 is a free R-module. Then A is a free $\Lambda/(\pi^d)$-module.

<u>Proof</u>: Let $\alpha_1,\ldots,\alpha_k \in A$ be chosen so that the cosets mod πA

are a basis of $A^0 = A/\pi A$. Then $\{\alpha_i\}$ form a basis of A. We see this, for example, by induction on d. Note that πA is elementary of degree $d - 1$ and $\{\pi\alpha_i\}$ define a basis of $(\pi A)^0 = \pi A/\pi^2 A$. Therefore $\{\pi\alpha_i\}$ is a basis of πA, by induction. Now any relation $\Sigma\lambda_i\alpha_i$, $\lambda_i \in \Lambda$, defines a relation $\Sigma\lambda_i(\pi\alpha_i) = 0$ and so every λ_i is divisible by π^{d-1}. Set $\lambda_i = \pi^{d-1}\mu_i$ and we have that $\Sigma\mu_i\alpha_i \equiv 0 \bmod \pi A$. This implies $\mu_i \equiv 0 \bmod \pi$ since $\{\alpha_i\}$ are linearly independent mod πA. Similarly, if $\alpha \in A$, then $\alpha \equiv \Sigma\lambda_i\alpha_i \bmod \pi A$ for some $\lambda_i \in \Lambda$, since $\{\alpha_i\}$ generate $A \bmod \pi A$. Therefore $\alpha - \Sigma\lambda_i\alpha_i \in \pi A$ and, by induction, there exist $\mu_i \in \nabla$ so that $\alpha - \Sigma\lambda_i\alpha_i = \Sigma\mu_i(\mu\alpha_i)$.

We will now extend the notion of degree to general π-primary Λ-modules, by saying degree $A = d$ if $\pi^d A = 0$ but $\pi^{d-1}A \neq 0$.

Lemma 15.2: A finitely generated π-primary Λ-module A of degree $\leq d$ is projective over $\Lambda/(\pi^d)$ only if it is elementary of degree d and A^0 is projective over R. If π is Dedekind, this criterion is sufficient for A to be projective.

Proof: A is projective if and only if, for some π-primary Λ-module B with $\pi^d B = 0$, $A \oplus B$ is free over $\Lambda/(\pi^d)$. Then $\pi(A \oplus B) = \operatorname{Ker} \pi^{d-1}$ and so $\pi A \oplus \pi B = \operatorname{Ker} \pi^{d-1}$ (on A) $\oplus \operatorname{Ker} \pi^{d-1}$ (on B) which implies $\pi A = \operatorname{Ker} \pi^{d-1}$. Furthermore $A^0 \oplus B^0 = (A \oplus B)^0$ which is free over R.

Conversely suppose A is elementary of degree d and A^0 is projective over R. Then for some finitely generated R-module M $A^0 \oplus M$ is free over R. By Lemma 14.1 there is an elementary π-primary module B of degree d with $B^0 \approx M$. Then $A \oplus B$ is also elementary of degree d and $(A \oplus B)^0 \approx A^0 \oplus M$ is free over R. By Lemma 15.1 $A \oplus B$ is free over $\Lambda/(\pi^d)$, which implies A is projective.

Lemma 15.3: Let A be an elementary π-primary Λ-module of degree d, π Dedekind and B any π-primary Λ-module with $\pi^d B = 0$. If $\phi: A^0 \to B^0$ is an R-module homomorphism, then there exists a Λ-module homomorphism $\Phi: A \to B$ which induces ϕ.

Proof: By Lemma 15.2, A is projective over $\Lambda/(\pi^d)$. Thus the conjugate $A \to A/\pi A \overset{\phi}{\to} B/\pi B$ can be lifted to a homomorphism Φ which obviously has the required property.

Lemma 15.4: Let A, B be π-primary Λ-modules, $\Phi: A \to B$ a homomorphism, $\phi: A_{d-1} \to B_{d-1}$ the induced homomorphism. If A is elementary of degree d, then Φ is injective if and only if ϕ is injective. If B is elementary of degree d, then Φ is surjective if and only if ϕ is surjective.

Proof: Note that ϕ is equivalent to $\Phi/\pi^{d-1}A: \pi^{d-1}A \to \pi^{d-1}B$. Clearly injectivity or surjectivity of Φ implies the same for ϕ.

Suppose ϕ is injective and A elementary. If $\Phi(\alpha) = 0$, choose k so that $\pi^k\alpha \neq 0$ and $\pi^{k+1}\alpha = 0$. Then $\pi^k\alpha \in \operatorname{Ker} \pi = \pi^{d-1}A$ and so $\phi(\pi^k\alpha) = 0$.

Suppose ϕ is surjective and B elementary. Assume $\pi^{k+1}B \subseteq \Phi(A)$. If $\beta \in \pi^k B$, write $\beta = \pi^k\beta'$. Since ϕ is surjective

$\pi^{d-1}\beta' = \Phi(\pi^{d-1}\alpha')$ for some $\alpha' \in A$ and so $\pi^k\beta' - \Phi(\pi^k\alpha') \in \operatorname{Ker} \pi^{d-1-k} = \pi^{k+1}B \subseteq \Phi(A)$. Thus $\beta = \pi^k\beta' \in \Phi(A)$.

Lemma 15.5: Let A, B be elementary π-primary Λ-modules of the same degree, with π Dedekind. Then $A \approx B$ if and only if $A^0 \approx B^0$.

Proof: This follows directly from Lemmas 15.3 and 15.4.

§16. Completion of proof

We can now prove Theorem 10.1. We proceed by induction on k. Given a family of exact sequences:

$$\{0 \to A_{i+1} \to A_i \to A^i \to A^{i+1} \to 0\},$$

where $A_{k+1} = A^{k+1} = 0$, we first construct a new family $\{0 \to B_{i+1} \to B_i \to B^i \to B^{i+1} \to 0\}$ with $B_k = B^k = 0$, as in §11. By induction we may assume there exists a π-primary Λ-module B with $\pi^k B = 0$ whose π-primary sequences are isomorphic to $\{0 \to B_{i+1} \to B_i \to B^i \to B^{i+1} \to 0\}$. Moreover we have a homomorphism $\phi: A_k \to B^0$ as in Proposition 12.1. By Lemma 14.1, let C be an elementary π-primary Λ-module of degree $k = 1$, with $C^0 \approx A_k$. Then πC is, of course, elementary of degree k and $(\pi C)^0 \approx C^0$. Let $\Phi: \pi C \to B$ be a homomorphism whose induced map $(\pi C)^0 \approx C^0 \approx A_k \to B^0$ corresponds to ϕ --by Lemma 15.3. We then define $A = (C \oplus B)/D$, where D is the submodule consisting of all elements of the form $(c, -\Phi(c))$, $c \in \pi C$.

Clearly $\pi^{k+1}A = 0$. We may consider $B \subseteq A$ by means of the injection $b \to (0, b)$, and, as such, coincides with Ker π^k. In fact, if (c, b) represents an element of Ker π^k, then $\Phi(\pi^k c) = -\pi^k b = 0$. But Φ is injective, by Lemma 15.4, since ϕ is injective.

To show that the module A has π-primary sequences isomorphic to the $\{0 \to A_{i+1} \to A_i \to A^i \to A^{i+1} \to 0\}$, it suffices, by Proposition 12.1 and Corollary 12.2 to verify that ϕ corresponds to the map $A/B \to B/\pi B = B^0$, induced by multiplication by π, under some isomorphic $A_k \approx A/B$. Define $\sigma\colon C^0 = C/\pi C \approx A/B$ by the formula $\sigma(c) = (c, 0)$. This map is well defined, since $(\pi c, 0) = (0, \Phi(\pi c))$ in A, and bijectivity is a straightforward consequence of the definition of A. Now $\Phi\colon \pi C \to B$ is defined so that the induced map $(\pi C)^0 \to B^0$ corresponds to ϕ, under the isomorphism $(\pi C)^0 \approx C^0 \approx A_k$. On the other hand, the composite $C^0 \overset{\sigma}{\to} A/B \overset{\pi}{\to} B/\pi B = B^0$ is given by: $c \mapsto (c, 0) \to (\pi c, 0) = (o, \Phi(\pi c)) = \Phi(\pi c) \in B$, regarded as included in A. This completes the proof of Theorem 10.1.

§17. Classification of Π-primary modules

We turn now to the question of classification of π-primary Λ-modules by their π-primary sequences.

Theorem 17.1: Let A, B be π-primary Λ-modules with π-only torsion of degree ≤ 3. Given homomorphisms $\phi_i\colon A_i \to B_i$, $\phi^i\colon A^i \to B^i$, there exists a homomorphism $\phi\colon A \to B$ which induces $\{\phi_i\}$.

Proof: Let $A(i)$, $B(i)$ denote Kernel π^i in A, B, respectively.

Define $\phi(1): A(1) \to B(1)$ to coincide with ϕ_0, since $A_0 = A(1)$, $B_0 = B(1)$. Note that $\phi(1)(\pi A(2)) \subseteq \pi B(2)$, $\phi(1)(\pi^2 A) \subseteq \pi^2 B$ and any extension of $\phi(1)$ over A will automatically induce the given $\{\phi_i\}$.

Notice that there is a commutative diagram:

$$(*) \qquad \begin{array}{ccccccccc} 0 & \to & A(2)/\pi A(2) & \overset{\theta}{\to} & A^0 \oplus A_1 & \overset{\rho}{\to} & A^1 & \to & 0 \\ & & \downarrow\bar{\phi} & & \downarrow\phi^0 \ \phi_1 & & \downarrow\phi^1 & & \\ 0 & \to & B(2)/\pi B(2) & \overset{\theta'}{\to} & B^0 \oplus B_1 & \overset{\rho'}{\to} & B^1 & \to & 0 \end{array}$$

The map ρ is the difference of the maps $A^0 \to A^1$, $A_1 \to A^1$ from the π-primary sequences of A. θ is the sum of the maps induced by inclusions. Similarly for θ^1, ρ^1, of course. We leave to the reader the exercise of showing that the rows are exact. The map $\bar{\phi}$ is induced from $\phi^0 \oplus \phi_1$ by commutativity of the diagram.

Also notice that the following diagrams are commutative:

$$(**) \qquad \begin{array}{ccccc} A(1) & \overset{i}{\to} & A(2)/\pi A(2) & \overset{\pi}{\to} & A(1) \\ \downarrow\phi(1) & ⓐ & \downarrow\bar{\phi} & ⓑ & \downarrow\phi(1) \\ B(1) & \overset{i'}{\to} & B(2)/\pi B(2) & \overset{\pi'}{\to} & B(1) \end{array} \qquad \begin{array}{ccc} A_2 & \overset{\tilde{\pi}}{\to} & \dfrac{A(2)}{\pi A(2)} \\ \phi_2\downarrow & ⓒ & \downarrow\bar{\phi} \\ B_2 & \overset{\tilde{\pi}'}{\to} & \dfrac{B(2)}{\pi B(2)} \end{array}$$

The maps i, i' are induced by inclusion π, π' and $\tilde{\pi}$, $\tilde{\pi}'$ are induced by multiplication by π. Commutativity of (**) follows from (*) and commutativity of

$$\begin{array}{ccccccc} A_0 = A(1) & \xrightarrow{\theta \cdot i} & A^0 \oplus A_1 & \xrightarrow{\pi \cdot \theta} & A_0 = A(1) \\ \downarrow\phi_0 \quad \downarrow\phi(1) & & \downarrow\phi^0 \oplus \phi_1 & & \downarrow\phi_0 \quad \downarrow\phi(1) \\ B_0 = B(1) & \xrightarrow{\theta' \cdot i'} & B^0 \oplus B_1 & \xrightarrow{\pi' \cdot \theta'} & B_0 = B(1) \end{array} \qquad \begin{array}{ccc} A_2 & \xrightarrow{\theta \cdot \tilde{\pi}} & A^0 \oplus A_1 \\ \phi_2\downarrow & & \downarrow\phi^0 \oplus \phi_1 \\ B_2 & \xrightarrow{\theta' \cdot \tilde{\pi}'} & B^0 \oplus B_1 \end{array}$$

since $\theta \cdot i$ is easily seen to be the sum of the map $\Delta_0: A_0 \to A^0$

from the 0-th π-primary sequence of A, and the zero map $A(1) \to A_1$, $\pi \cdot \theta$ coincides with projection on A_1 followed by the map $\pi_0: A_1 \to A_0$ from the 0-th π-primary sequence of A and $\theta \cdot \tilde{\pi}$ is zero plus the map $\pi_1: A_2 \to A_1$ from the 1-st π-primary sequence of A.

We now extend $\phi(1)$ over $A(2)$. By Lemma 14.1, there exists an elementary π-primary $\widetilde{A(2)}$ of degree 2 such that $(\widetilde{A(2)})_1 \approx A_1 = A(2)_1$. In fact, we may assume $\widetilde{A(2)} \subseteq A(2)$ so that $(\widetilde{A(2)})_1 \to A(2)_1$, induced by the inclusion is an isomorphism. To construct such an injection, lift the given isomorphism $(\widetilde{A(2)})^0 = \widetilde{A(2)}_1 \approx A(2)_1$ to a homomorphism $(\widetilde{A(2)})^0 \to A(2)^0$ --since $\widetilde{A(2)}^0$ is a projective R-module and $A(2)^0 \to A(2)_1$, induced by inclusion, is an epimorphism. Now apply Lemmas 15.3 and 15.4. Note that $A(2) = \widetilde{A(2)} + A(1)$ and $\widetilde{A(2)} \cap A(1) = \pi\widetilde{A(2)} = \pi A(2)$. Since $\widetilde{A(2)}$ is projective over $\Lambda/(\pi^2)$, by Lemma 15.2, there exists a map $\widetilde{\phi(2)}: \widetilde{A(2)} \to B(2)$ satisfying the commutative diagram:

$$\widetilde{A(2)} \subseteq A(2) \to A(2)/\pi A(2)$$
$$\phi(2) \searrow \qquad\qquad \downarrow \overline{\phi}$$
$$B(2) \to B(2)/\pi B(2)$$

We will show that $\widetilde{\phi(2)}$ agrees with $\phi(1)$ on $\widetilde{A(2)} \cap A(1) = \pi\widetilde{A(2)}$.

Suppose $\alpha \in \widetilde{A(2)}$, then $\widetilde{\phi(2)}(\alpha) \equiv \overline{\phi}([\alpha]) \mod \pi B(2)$, where $[\alpha]$ denotes coset mod $\pi A(2)$. Therefore $\widetilde{\phi(2)}(\pi\alpha) = \pi\widetilde{\phi(2)}\alpha = \pi'\overline{\phi}([\alpha])$, where the last π' denotes the map $B(2)/\pi B(2) \to B(1)$ of $(**,b)$, and $\pi'\overline{\phi}[\alpha] = \phi(1)(\pi\alpha)$ from the commutativity of $(**, b)$.

Now we may construct $\phi(\overset{.}{2}): A(2) \to B(2)$ which agrees with $\phi(2)$

on $\tilde{A}(2)$ and $\phi(1)$ on $A(1)$, since $A(2) = \tilde{A}(2) + A(1)$. It follows from the commutativity of (**, a), if $\alpha \varepsilon A(1)$, and definition of $\tilde{\Phi}(2)$, if $\alpha \varepsilon \widetilde{A(2)}$, that $\phi(2)(\alpha) \equiv \bar{\phi}([\alpha]) \bmod \pi B(2)$, for any $\alpha \varepsilon A(2)$.

We now extend $\phi(2)$ to a map $\phi\colon A \to B$. By the same argument as above, there exists an elementary π-primary $\tilde{A}$ of degree 3 such that $\tilde{A} \subseteq A$ inducing an isomorphism $\tilde{A}_2 \approx A_2$. Now construct $\tilde{\phi}\colon \tilde{A} \to B$ to satisfy the commutative diagram:

$$\begin{array}{ccc} \tilde{A} \subseteq A & \to & A_2 \\ \tilde{\phi} \searrow & & \downarrow \phi_2 \\ B & \to & B_2 \end{array}$$

where the horizontal maps are quotient maps. Now $A = \tilde{A} + A(2)$ and $\tilde{A} \cap A(2) = \pi\tilde{A}$. We will show that $\tilde{\phi}$ may be chosen to agree with $\phi(2)$ on $\pi\tilde{A}$. If $\alpha \varepsilon \tilde{A}$, then $\tilde{\phi}(\alpha) \equiv \phi_2([\alpha]) \bmod B(2)$ where $[\alpha]$ is the coset of α in A_2. So $\tilde{\phi}(\pi\alpha) = \pi\tilde{\phi}(\alpha) = \tilde{\pi}'\phi_2[\alpha] \bmod \pi B/2)$, where $\tilde{\pi}'$ is the map of (**, c). Now $\tilde{\pi}'\phi_2[\alpha] = \bar{\phi}\tilde{\pi}[\alpha]$, by (**, c), and $\bar{\phi}\tilde{\pi}[\alpha] = \bar{\phi}[\alpha\pi]$, where $[\pi\alpha]$ is the coset of $\pi\alpha$, mod $\pi A(2)$. Thus $\tilde{\phi}(\pi\alpha) \equiv \bar{\phi}[\pi\alpha] \bmod \pi B(2)$. On the other hand $\phi(2)(\beta) \equiv \bar{\phi}[\beta] \bmod \pi B(2)$, for any $\beta \varepsilon A(2)$. Thus $\tilde{\phi} - \phi(2)$ defines a map $\pi\tilde{A} \to \pi B(2)$. By projectivity of $\tilde{A}$, we may define $\psi\colon \tilde{A} \to B(2)$ satisfying:

$$\begin{array}{ccc} \tilde{A} & \overset{\pi}{\to} & \Pi\tilde{A} \\ \psi\downarrow & & \downarrow\phi - \phi(2) \\ B(2) & \overset{\pi}{\to} & \Pi B(2) \end{array}$$

If $\alpha \varepsilon \tilde{A}$, then $\psi(\pi\alpha) = \pi\psi(\alpha) = (\tilde{\phi} - \phi(2))(\pi\alpha)$, so $\psi = \tilde{\phi} - \phi(2)$

on $\pi\tilde{A}$. Therefore $\tilde{\phi} - \psi: \tilde{A} \to B$ agrees with $\phi(2)$ on $A(2)$, and so induces a map $\phi: A \to B$ which extends $\phi(2)$ as desired. This completes the proof of Theorem 17.1.

Corollary 17.2: Let A, B be finitely generated π-primary Λ-modules with π-only torsion of degree ≤ 3. If π is Dedekind, then $A \approx B$ if and only if the π-primary sequences of A, B are isomorphic.

Proof: By theorem 17.1, there exists a homomorphism $\phi: A \to B$ whose induced maps $\phi_i: A_i \to B_i$ are isomorphisms. Since $\{A_i\}$ are quotients of successive members of a filtration of A, it follows from repeated application of the 5-lemma that ϕ is itself an isomorphism.

Corollary 17.3: Let A be a π-primary Λ-module with π-only torsion and degree ≤ 2. If π is Dedekind, then the isomorphism class of A is determined by the pair $\pi A \subseteq \text{Ker } \pi$ of R-modules. Given any pair $A_1 \subseteq A_0$ of finitely generated R-torsion free R-modules, there exists a module A, as above, with $(\text{Ker } \pi, \pi A) \approx (A_0, A_1)$.

Note that the pairs (A_0, A_1) are classified by the rank and ideal class of A_0 and the rank, ideal class and invariant factors of A_0/A_1 (see e.g. [CR: 22, esp. p.154, Ex. 6]). This gives a complete and effective classification of the modules A of Corollary 17.3.

The proof of Corollary 17.3 follows from Corollary 17.2, since the

π-primary sequence of A reduces to the inclusion $\pi A \subseteq \mathrm{Ker}\ \pi$, and Theorem 10.1.

Corollary 17.4: If A is as in Corollary 17.3 and π Dedekind, then A splits into a direct sum $A' \oplus A''$, where A', A'' are uniquely determined homogeneous modules of degree 1, 2, respectively.

Proof: By the classification of pairs of R-torsion free R-modules [CR: §22], we may decompose $\mathrm{Ker}\ \pi = B' \oplus B''$, with $\pi A \subseteq B''$ and $B''/\pi A$ an R-torsion module. If we define $A' = B'$ and A'' the module associated with the pair $\pi A \subseteq B''$ by Corollary 17.3, then $A \approx A' \oplus A''$, by Corollary 17.3 again. Furthermore, A'' is homogeneous of degree 2, by Proposition 13.1.

To see the uniqueness of A', A'' notice that $A' \approx \mathrm{Ker}\ \pi/\pi A$, A'' is determined, according to Corollary 17.3, by the pair $\pi A \subseteq \widehat{\pi A}$, where $\widehat{\pi A} = \{\alpha \in \mathrm{Ker}\ \pi\colon \lambda\alpha \in \pi A$ for some $\lambda \in R - \{0\}\}$.

It is not true, however, that every π-prime A of degree 3 is a direct sum of homogeneous modules. Consider, for example, A the submodule of $V = \Lambda/(\pi^3) \oplus \Lambda/(\pi^3)$ generated by the elements $(\bar{\sigma}^2, 0)$ $(0, \pi\bar{\sigma})$, (π^2, π^2) where $\bar{\sigma} \in \Lambda$ determines an element $\sigma \in R$ which is nonzero and not a unit.

Proposition 17.5: A is not a sum of homogeneous modules.

Proof: Consider the injections $A_2 \xrightarrow{\pi_1} A_1 \xrightarrow{\pi_0} A_0 = \mathrm{Ker}\ \pi \subseteq \pi^2 V$. If we identify $\pi^2 V \approx R \oplus R$, then the images of A_2, A_1, and A_0 are described as follows: A_0 generated by $(0, \sigma)$, $(1, 1)$; A_1 generated by $(\sigma^2, 0)$, $(0, \sigma)$; A_2 generated by $(\sigma^2, 0)$. If A were a sum of homogeneous modules, then, by consideration of the ranks of $\{A_i\}$, we would have $A = A' \oplus A''$, where A' has degree 2, A'' degree 3. We would then have splittings $A_i = (A')_i \oplus (A'')_i$ consistent with the injections above. Obviously $(A'')_2 = A_2$ is generated by $(\sigma^2, 0)$. Since $(A'')_i$ is characterized as the set of all $\alpha \in A_i$ such that $\lambda\alpha \in A''_2$ for some non-zero $\lambda \in R$ --by Proposition 13.1--we compute easily that $A''_1 = A''_2$ and A''_0 is free generated by $(\sigma, 0)$. Since the quotient A''_0/A''_1 has order (σ) and A_0/A_1 has order (σ^2), it follows that A'_0/A'_1 has order (σ). Now A'_0, as a complementary summand of A''_0 in A_0 must contain an element $\alpha = (1, 1) + \lambda(\sigma, 0)$, for some $\lambda \in R$, which clearly must then generate A'_0. Therefore A'_1 will have to have $\sigma\alpha = (\sigma, \sigma) - \lambda(\sigma^2, 0)$ as a generator. But A_1 does not contain the element (σ, σ), so cannot contain $\sigma\alpha$. This proves the proposition.

§18. Classification fails in degree 4

We now show, by an example, that the classification of π-primary Λ-modules by their π-primary sequences is not possible, for degree > 3, if R is not a field.

Let $S = \Lambda/(\pi^4)$ and $\sigma \in S$ an element whose reduction in $R = \Lambda/(\pi) = S/\pi^3 S$ is non-zero and a non-unit. Consider the ideals I, I' of S defined by

$$I = (\sigma^4, \sigma^3\pi, \sigma\pi^2, \pi^3)$$
$$I' = (\sigma^4 + \sigma^2\pi, \sigma^3\pi, \sigma\pi^2, \pi^3)$$

<u>Theorem 18.1</u>: I and I' are not isomorphic as π-modules, but their π-primary sequences are isomorphic..

<u>Proof</u>: We first show $I \not\approx I'$. Consider the localization S_π of S, at the prime ideal (π). It is easy to see that $I \approx I'$ if and only if $I' = \lambda I$, for some unit $\lambda \in S_\pi$, because $I_\pi = S_\pi = I'_\pi$. We investigate the possible values of such λ.

Since $\lambda(I \cap \pi^3 S) = I' \cap \pi^3 S$, we have $\lambda\pi^3 S = \pi^3 S$. It follows that λ is a unit of S, mod π. Dividing λ by this unit, it follows that we may assume λ has the form $\lambda = 1 + \lambda'\pi$, for some $\lambda' \in S_\pi$. But $\sigma\pi^2 \in I'$, so we can conclude $\lambda'\sigma\pi^3 \in I'$ and, therefore, $\lambda'\sigma\pi^3 \in \pi^3 S$. Thus $\lambda'\sigma \in S \bmod \pi$ and we may write $\lambda' = \frac{\alpha}{\sigma} + \pi\lambda''$, for $\alpha \in S$, $\lambda'' \in S_\pi$. Now $\lambda = 1 + \frac{\alpha}{\sigma}\pi \bmod \pi^2$ and so $\lambda I = (\sigma^4 + \alpha\sigma^3\pi, \sigma^3\pi) = (\sigma^4, \sigma^3\pi) \bmod \pi^2$ while $I' = (\sigma^4 + \sigma^2\pi, \sigma^3\pi) \bmod \pi^2$. These are different since, for example, $\sigma^4 \notin I' \bmod \pi^2$.

The computation of the π-primary sequences of I and I' is straightforward and we omit the details. For both I and I' we obtain obvious isomorphisms: $A_0 \approx R$, $A_1 \approx R$, $A_2 \approx R$, $A_3 \approx R$, $A^0 \approx R \oplus R/\sigma \oplus R/\sigma^2 \oplus R/\sigma$, $A^1 \approx R \oplus R/\sigma \oplus R/\sigma^2$, $A^2 \approx R \oplus R/\sigma$, $A^3 \approx R$. Under these isomorphisms the π-primary sequences are:

$$0 \to A_1 \xrightarrow{\sigma} A_0 \xrightarrow{(0,0,0,1)} A^0 \xrightarrow{(1,1,1,0)} A^1 \to 0$$
$$0 \to A_2 \xrightarrow{\sigma^2} A_1 \xrightarrow{(0,0,1)} A^1 \xrightarrow{(1,1,0)} A^2 \to 0$$
$$0 \to A_3 \xrightarrow{\sigma} A_2 \xrightarrow{(0,1)} A^2 \xrightarrow{(1,0)} A^3 \to 0$$
$$0 \to A_3 \xrightarrow{1} A^3 \to 0$$

§19. Product structure on Π-primary modules

We now introduce the Blanchfield pairing into our considerations. If A is a finitely generated Λ-module, where Λ is a unique factorization domain with an involution $\alpha \mapsto \bar{\alpha}$, we will say that a function:

$$<,>: A \times A \to Q(\Lambda)/\Lambda = S(\Lambda), \quad Q(\Lambda) = \text{quotient field of } \Lambda$$

is an ϵ-pairing ($\epsilon = \pm 1$) if it satisfies:

i) conjugate linearity: $<\lambda\alpha, \beta> = <\alpha, \bar{\lambda}\beta> = \lambda<\alpha, \beta>$ for any $\lambda \in \Lambda$; $\alpha, \beta \in A$.

ii) ϵ-Hermitian: $<\alpha, \beta> = \epsilon\overline{<\beta, \alpha>}$ for any $\alpha, \beta \in A$

$<,>$ defines an adjoint homomorphism $\psi: \bar{A} \to \mathrm{Hom}_\Lambda(A, S(\Lambda))$, by $\psi(\alpha) \cdot \beta = <\beta, \alpha> = \epsilon\overline{<\alpha, \beta>}$, where $\bar{A}$ is defined to be A with a new Λ-module structure: $(\lambda\alpha)_{new} = (\bar{\lambda}\alpha)_{old}$.

We say $<,>$ is non-singular if ψ is an isomorphism onto, non-degenerate if ψ is injective (see [L: §4, 5]).

Lemma 19.1: If $<,>$ is non-degenerate and A is π-primary, then A has π-only torsion, and π may be chosen to satisfy $\pi = \bar{\pi}$.

Proof: Suppose $\lambda\alpha = 0$, where λ is not divisible by π. Then $\pi^m\alpha = 0$ for some m, and $\pi^m<\alpha, \beta> = 0 = \lambda<\alpha, \beta>$ for any $\beta \in A$. If $<\alpha, \beta> \in S(\Lambda)$ is represented by a quotient $\sigma/\tau \in Q(\Lambda)$, then both $\pi^m\sigma$ and $\lambda\sigma$ are divisible by τ. Since π and λ are assumed relatively prime and Λ is a unique factorization domain, it follows that σ is divisible by τ and so $<\alpha, \beta> = 0$. Since this holds for all $\beta \in A$, and $<,>$ is non-degenerate, we conclude $\alpha = 0$.

For the last part of the lemma, suppose $\pi^m A = 0$, then $\langle \bar{\pi}^m \alpha, \beta \rangle = \langle \alpha, \pi^m \beta \rangle = 0$, for every $\alpha, \beta \in A$. Since $\langle , \rangle$ is non-degenerate, $\bar{\pi}^m A = 0$. Thus $\bar{\pi}^m$ is divisible by π.

We introduce the following notation. If A is a π-primary Λ-module and $B \subseteq A$ submodule, define $\hat{B}$ to consist of all $\alpha \in A$ such that $\lambda\alpha \in B$, for some non-zero $\lambda \in \Lambda$ relatively prime to π. $\hat{B}$ is a submodule of A containing B and is the smallest such that $A/\hat{B}$ has π-only torsion. We may also define $\hat{B} = A \cap \Lambda_\pi B$, where Λ_π is the localization of Λ at the prime ideal (π).

If B is a submodule of A and $\langle , \rangle$ is an ϵ-form on A, then the <u>annihilator of B</u>, denoted B^0, is the submodule of all α such that $\langle \alpha, \beta \rangle = 0$, for all $\beta \in B$.

<u>Lemma 19.2</u>: Let A be a finitely generated π-primary Λ-module with π-only torsion, $B = \hat{B}$ a submodule of A, and $\phi: B \to S(\Lambda)$ a homomorphism. Then ϕ extends over A provided:

i) $\Lambda = Z[t, t^{-1}]$ and $\pi(1) = \pm 1$, or

ii) π is Dedekind.

<u>Proof</u>: In case (i) it follows from [I: 3.2] that $\mathrm{Ext}^1_\Lambda(C, S(\Lambda)) \approx \mathrm{Ext}^2_\Lambda(C, \Lambda) = 0$ for any Z-torsion free $Z[t, t^{-1}]$-module C of type K. Since A/B is Z-torsion free, by Corollary 9.4, the lemma follows.

In case (ii) let $B_i = B + \mathrm{Ker}\, \pi^i$ and assume ϕ has been extended over B_i --we show how to extend over B_{i+1}. Since B_{i+1}/B_i is R-torsion free and π is Dedekind, then, by Lemmas 14.1, 15.2, 15.3, and 15.4, there exists an elementary submodule P, of degree $i+1$, of B_{i+1} such that $P/\pi P \to B_{i+1}/B_i$ is an isomorphism. Therefore $B_{i+1} = P + B_i$ and $P \cap B_i = \pi P$. Extend $\phi|\pi P$ to $\phi': P \to S(\Lambda)$ giving a commutative diagram:

$$\begin{array}{ccc} P & \xrightarrow{\phi'} & S(\Lambda) \\ \downarrow\pi & & \downarrow\pi \\ \pi P & \xrightarrow{\phi} & S(\Lambda) \end{array}$$

Clearly $\phi'|\pi P = \phi$ and so we can piece ϕ and ϕ' together to give the desired extension over B_{i+1}.

We will say the prime element $\pi \in \Lambda$ is <u>workable</u> if (i) or (ii) of Lemma 19.2 is satisfied.

We can reformulate Lemma 19.2 in the following useful way:

<u>Lemma 19.2'</u>: If Λ and A, as in 19.2, with π workable, $B \subseteq A$ any submodule, and $\phi: B \to S(\Lambda)$, then for some $\lambda \in \Lambda$, relatively prime to π, $\lambda\phi$ extends over A.

<u>Proof</u>: Λ is noetherian if π is workable, and so, for some λ relatively prime to π, $\lambda \hat{B} \subseteq B$. Cleraly $\lambda\phi$ extends over $\hat{B}$, by defining $\phi'(\alpha) = \phi(\lambda\alpha)$ for any $\alpha \in B$. Now $\hat{\hat{B}} = \hat{B}$, so we can apply 19.2 to extend ϕ' over A.

<u>Lemma 19.3</u>: If $<,>$ is an ϵ-form on a π-primary Λ-module A and $B \subseteq A$ submodule, then $\widehat{B^0} = B^0$.

Proof: Suppose $\lambda\alpha \in B^0$, $\lambda \notin (\pi)$. Then $\lambda\langle\alpha, \beta\rangle = \langle\lambda\alpha, \beta\rangle = 0$, for all $\beta \in B$. But $\pi^m\langle\alpha, \beta\rangle = 0$ also, for some m. Since λ and π^m are relatively prime in Λ, we conclude $\langle\alpha, \beta\rangle = 0$.

Lemma 19.4: Let $\langle,\rangle$ be a non-singular ϵ-form on a finitely generated π-primary Λ-module A, and B, C submodules. Then if π is workable:

a) $(B \cap C)^0 = \widehat{B^0 + C^0}$

b) $(B + C)^0 = B^0 \cap C^0$

Proof: (b) is easy and does not require non-singularity or π workable. The inclusion $(B \cap C)^0 \subseteq \widehat{B^0 + C^0}$ is also clear, using 19.3, and also does not require non-singularity or π workable.

Suppose $\alpha \in (B \cap C)^0$. Then $\langle\cdot, \alpha\rangle$ defines a homomorphism $A/B\cap C \to S(\Lambda)$. Using the isomorphism $B/B \cap C \approx \frac{B+C}{C}$, $\langle\ , \alpha\rangle|B$ corresponds to a homomorphism $\phi: B + C \to S(\Lambda)$ such that $\phi|C = 0$. By Lemma 19.2', $\lambda\phi$ extends to ϕ' over A, for some λ relatively prime to π. Since $\langle,\rangle$ is non-singular, there exists $\gamma \in A$ such that $\langle\cdot, \gamma\rangle = \phi'$. Furthermore $\gamma \in C^0$, since $\phi'|C = \lambda\phi|C = 0$. By construction of ϕ', $\langle\cdot, \gamma\rangle = \lambda\langle\cdot, \alpha\rangle$ on B; therefore $\bar{\lambda}\alpha - \gamma \in B^0$. In other words, $\bar{\lambda}\alpha \in B^0 + C^0$ and so $\alpha \in \widehat{B^0 + C^0}$, as was to be proved.

Lemma 19.5: Let $\langle,\rangle$ be a non-singular ϵ-form on the finitely generated π-primary Λ-module A, with π workable. Then

$(\text{Ker } \pi^i)^0 = \widehat{\pi^i A}$

$(\pi^i A)^0 = \text{Ker } \pi^i$

Proof: The inclusions $\pi^i A \subset (\text{Ker } \pi^i)^0$, and $\text{Ker } \pi^i \subseteq (\pi^i A)^0$ are clear, since, if $\alpha \in \text{Ker } \pi^i = \text{Ker } \bar{\pi}^i$ by 19.1 and $\beta \in A$ $\langle\alpha, \pi^i\beta\rangle = \langle\bar{\pi}^i\alpha, \beta\rangle = 0$. By 19.3 we obtain $\widehat{\pi^i A} \subseteq (\text{Ker } \pi^i)^0$. Furthermore, $(\pi^i A)^0 \subseteq \text{Ker } \pi^i$ since $\langle\alpha, \pi^i A\rangle = 0$ implies $\langle\bar{\pi}^i\alpha, A\rangle = 0$ and so $\alpha \in \text{Ker}\bar{\pi}^i = \text{Ker } \pi^i$.

It remains to prove $(\text{Ker } \pi^i)^0 \subseteq \widehat{\pi^i A}$. Suppose $\alpha \in (\text{Ker } \pi^i)^0$. Then $\langle\cdot, \alpha\rangle$ defines a homomorphism $A/\text{Ker } \pi^i \to S(\Lambda)$ which corresponds to a homomorphism $\phi\colon \pi^i A \to S(\Lambda)$ satisfying $\phi(\pi^i\beta) = \langle\beta, \alpha\rangle$, for any $\beta \in A$. By Lemma 19.2', $\lambda\phi$ extends to $\phi'\colon A \to S(\Lambda)$, for some λ not divisible by π. Since $\langle,\rangle$ is non-singular, there exists $\gamma \in A$ such that $\langle\cdot, \gamma\rangle = \phi'$. Thus $\langle\cdot, \gamma\rangle = \lambda\phi$ on $\pi^i A$, so for any $\beta \in A$,
$\langle\beta, \bar{\lambda}\alpha\rangle = \lambda\langle\beta, \alpha\rangle = \lambda\phi(\pi^i\beta) = \langle\pi^i\beta, \alpha\rangle = \langle\beta, \bar{\pi}^i\gamma\rangle$ since $\langle,\rangle$ is non-degenerate $\bar{\lambda}\alpha \in \bar{\pi}^i A = \pi^i A$ and so $\alpha \in \widehat{\pi^i A}$.

Lemma 19.6: Let $\langle,\rangle$ be an ϵ-form on a π-primary Λ-module A with π-only torsion. Let B be a submodule of A on which $\langle,\rangle$ is zero, i.e., $\langle\alpha, \beta\rangle = 0$ for any $\alpha, \beta \in B$. Then $\langle,\rangle$ induces an ϵ-pairing on $B^0/\hat{B}$, which is non-singular if $\langle,\rangle$ is non-singular, π is workable, and $(B^0)^0 = \hat{B}$.

Proof: Since $B \subseteq (B^0)^0$, we have $\hat{B} \subseteq \widehat{(B^0)^0} = (B^0)^0$, by 19.3. Thus the induced pairing $[,]$ is well defined. The condition $(B^0)^0 = \hat{B}$ is clearly equivalent to non-degeneracy of $[,]$. To prove non-singularity, suppose $\phi\colon B^0/\hat{B} \to S(\Lambda)$. Since $\widehat{B^0} = B^0$ by 19.3, ϕ extends to a homomorphism $\phi'\colon A \to S(\Lambda)$ --of course, $\phi'(\hat{B}) = 0$.

Since $<,>$ is non-singular, $\phi' = <\cdot, \alpha>$ for some $\alpha \in A$ --clearly $\alpha \in (\hat{B})^0 = B^0$, and so represents $\alpha' \in B^0/\hat{B}$. But this means $\phi = [\cdot, \alpha']$, which proves $[,]$ is non-singular.

§20. Classification of product structure

Let A be a finitely generated π-primary Λ-module with a non-degenerate ϵ-form $<,>$. Consider the π-primary sequences:

$$0 \to A_{i+1} \xrightarrow{\pi_i} A_i \to A^i \xrightarrow{\pi^i} A^{i+1} \to 0$$

Define $\Delta_i(A) = \text{Cok}\ \pi_i \approx \text{Ker}\ \pi^i$; thus

$$\Delta_i(A) = \frac{\pi^{i+1}A + \pi^i \text{Ker}\ \pi^{i+1}}{\pi^{i+1}A} \approx \frac{\pi^i \text{Ker}\ \pi^{i+1}}{\pi^{i+1}\text{Ker}\ \pi^{i+2}}$$

The ϵ-form $<,>$ induces an ϵ-form $<,>_i$ on $\Delta_i(A)$ as follows. Let $\alpha, \beta \in \text{Ker}\ \pi^{i+1}$ and define

$$<\pi^i\alpha, \pi^i\beta>_i = \pi^i<\alpha, \beta>$$

If $\alpha \in \pi\text{Ker}\ \pi^{i+2}$, say $\alpha = \pi\beta'$, then:

$$<\pi^i\alpha, \pi^i\beta>_i = \pi^i<\alpha, \beta> = \pi^{i+1}<\alpha', \beta> = <\alpha', \bar{\pi}^{i+1}\beta> = 0,$$

and so $<,>_i$ is well defined on $\Delta_i(A)$.

If $\tilde{\Delta}_i(A)$ denotes the R-torsion free quotient of $\Delta_i(A)$, then $<,>_i$ induces an ϵ-pairing on $\tilde{\Delta}_i(A)$, by 19.1, which we also denote $<,>_i$, with values in $R \xrightarrow[\subseteq]{1/\pi} S(\Lambda)$.

<u>Claim</u>: $<,>_i$ is non-degenerate on $\tilde{\Delta}_i(A)$, if π is workable.

Suppose $\alpha \in \pi^i\text{Ker}\ \pi^{i+1}$ and $<\alpha, \pi^i\beta>_i = <\alpha, \beta> = 0$ for all

$\beta \in \operatorname{Ker} \pi^{i+1}$. Then $\alpha \in (\operatorname{Ker} \pi^{i+1})^0 = \widehat{\pi^{i+1}A}$. So $\alpha \in \widehat{\pi^{i+1}A} \cap \pi^i \operatorname{Ker} \pi^{i+1} \subseteq \widehat{\pi^{i+1} \operatorname{Ker}} \pi^{i+2}$. But

$$\tilde{\Delta}_i(A) \approx \frac{\pi^i \operatorname{Ker} \pi^{i+1}}{\widehat{\pi^{i+1} \operatorname{Ker}} \pi^{i+2}}, \text{ and so } \alpha \text{ represents } 0 \text{ in } \tilde{\Delta}_i(A).$$

We will now show that the classification of π-primary Λ-modules with non-singular ϵ-forms can be derived from the classification of π-primary Λ-modules by using the derived ϵ-forms $\langle,\rangle_i$.

If R is a commutative ring with involution, we say $\mu \in R$ is a <u>half-unit</u> if $\mu + \bar{\mu} = 1$. Note that in any ring R with half-unit $\alpha = \epsilon\bar{\alpha}$ $(\epsilon = \pm 1)$ if and only if $\alpha = x + \epsilon\bar{x}$, for some $x \in R$ -- just set $x = \mu\alpha$.

If $\Lambda = Z[t, t^{-1}]$, π irreducible satisfying $\pi(1)$ odd and $\pi(t) = \pi(t^{-1})$, then $R = \Lambda/(\pi)$ inherits an involution from the standard involution $t \mapsto t^{-1}$ of Λ. This includes the case of interest in knot theory; R contains a half-unit. In fact $1 + \pi = \lambda$ satisfies $\lambda(t) = \lambda(t^{-1})$ and $\lambda(1)$ even. If we write $\lambda = \sum_i a_i t^i$, then $a_i = a_{-i}$ and a_0 is even. Setting $\mu = a_{0/2} + \sum_{i \geq 1} a_i t^i$, we see $1 + \pi = \lambda = \mu + \bar{\mu}$; thus μ represents a half-unit in R.

<u>Theorem 20.1</u>: Let A, B be finitely generated π-primary Λ-modules with nonsingular ϵ-forms $\langle,\rangle_A$ and $\langle,\rangle_B$. Suppose π is workable, $\pi = \bar{\pi}$ and $R = \Lambda/(\pi)$ contains a half-unit. Then A and B are isometric (i.e., isomorphic preserving the ϵ-forms) if and only if A and B are isomorphic via an isomorphism which induces isometries $\tilde{\Delta}_i(A) \approx \tilde{\Delta}_i(B)$ for every i.

Combining this with Corollary 17.2, we have

Corollary 20.2: Let A, B be finitely generated π-primary Λ-modules of degree $\leq$ 3, equipped with non-singular ϵ-forms. Assume π is Dedekind and R contains a half-unit. Then A and B are isometric if and only if there exist isomorphisms between their π-primary sequences, inducing isometries $\tilde{\Delta}_i(A) \approx \tilde{\Delta}_i(B)$.

Proof of Theorem 20.1: Let $\Phi: A \to B$ be an isomorphism which induces isometries $\tilde{\Delta}_i(A) \approx \tilde{\Delta}_i(B)$. We may formulate this as:

$(1)_i$ $\langle\alpha, \beta\rangle_A = \langle\Phi\alpha, \Phi\beta\rangle_B$ for any $\alpha \in \pi^i K_{i+1}(A)$, $\beta \in K_{i+1}(A)$ where $K_i(A) = \text{Ker}\ \pi^i$ in A. So $(1)_i$ holds for all i.

We first alter Φ so that $\langle\Phi\alpha, \Phi\beta\rangle_B = \langle\alpha, \beta\rangle_A$ for any $\alpha \in K_1(A)$. This is just the case $i = 0$ of:

$(2)_i$ $\langle\Phi\alpha, \Phi\beta\rangle_B = \langle\alpha, \beta\rangle_A$ if $\alpha \in \pi^i K_{i+1}(A)$, $\beta \in A$.

If A is of degree d, then $(2)_{d-1}$ holds, since this coincides with $(1)_{d-1}$. Assume inductively that $(2)_n$ holds--we alter Φ to achieve $(2)_{n-1}$.

Define $[\alpha, \beta] = \langle\Phi\alpha, \Phi\beta\rangle_B - \langle\alpha, \beta\rangle_A$. By $(2)_n$, $[,]$ is an ϵ-form on $A/\pi^n K_{n+1}(A)$, and, therefore, by 19.1, on $A/\widehat{\pi^n K_{n+1}}(A)$.

Claim: There exists $\mu \in \Lambda$ such that $\mu + \bar{\mu} \equiv 1 \bmod \pi^d$. By assumption, μ exists so that $\mu + \bar{\mu} \equiv 1 \bmod \pi$. Suppose $\mu + \bar{\mu} \equiv 1 \bmod \pi^i$ --then $\mu + \bar{\mu} = 1 + \pi^i\alpha$. Then $\pi^i\alpha = \pi^i\bar{\alpha}$, since $\pi = \bar{\pi}$.
Now, if $\mu' = \mu(1 - \pi^i\alpha)$, then
$\mu' + \bar{\mu}' = (\mu + \bar{\mu})(1 - \pi^i\alpha) = (1 + \pi^i\alpha)(1 - \pi^i\alpha) = 1 - \pi^{2i}\alpha^2 \equiv 1 \bmod \pi^{2i}$.

Now define a conjugate-linear pairing $\{\ ,\ \}$ on $A/\widehat{\pi^n K_{n+1}}(A)$ by $\{\alpha,\beta\} = \mu[\alpha,\beta]$. Then:

(3) $[\alpha,\beta] = \{\alpha,\beta\} + \epsilon\overline{\{\beta,\alpha\}}$

since $[\alpha,\beta] = (\mu+\bar{\mu})[\alpha,\beta] = \mu[\alpha,\beta] + \bar{\mu}[\alpha,\beta] = \{\alpha,\beta\} + \epsilon\overline{\mu[\beta,\alpha]}$

$= \{\alpha,\beta\} + \epsilon\overline{\{\beta,\alpha\}}$

Consider the homomorphism $A \to S(\Lambda)$, defined by $\{.,\alpha\}$, for some $\alpha \in A$. By non-singularity of $<,>_A$, there exists $\alpha' \in A$, such that $\{.,\alpha\} = <.,\alpha'>_A$. Since $\{\widehat{\pi^n K_{n+1}}(A),\alpha\} = 0$, we have

$\alpha' \in (\pi^n K_{n+1}(A))^o = (\pi^n A \cap K_1(A))^o = \widehat{(\pi^n A)^o + K_1(A)^o} = \widehat{K_n(A) + \pi A}$,

by (19.4),(19.5). The correspondence $\alpha \to \alpha'$ defines a homomorphism of Λ-modules.

$$h:\quad A/\widehat{\pi^n K_{n+1}}(A) \to \widehat{K_n(A) + \pi A}$$

with the defining property: $<\beta,h(\alpha)>_A = \{\beta,\alpha\}$, all $\alpha,\beta \in A$.

Define $\Psi: A \to A$ by $\Psi(\alpha) = \alpha + h(\alpha)$. We will prove:

(4) Ψ is an isomorphism

$(5)_n$ $<\Psi\alpha,\Psi\beta>_A = <\Phi\alpha,\Phi\beta>_B$, if $\alpha \in \pi^{n-1}K_n(A)$, $\beta \in A$

$(6)_i$ $<\Psi\alpha,\Psi\beta>_A = <\Phi\alpha,\Phi\beta>_B$, if $\alpha \in \pi^i K_{i+1}(A)$, $\beta \in k_{i+1}(A)$

To prove (4), it suffices to show h is nilpotent, i.e. $h^m(\alpha) = 0$, for large enough m. We first show:

$(7)_i$ $h(K_{i+1}(A)) \subseteq \widehat{K_i(A) + \pi A}$, for all i.

In fact if $\alpha \in K_{i+1}(A)$, $\beta \in \pi^i K_{i+1}(A)$, we have, by $(7)_i$

$$<\beta,h(\alpha)>_A = \{\beta,\alpha\} = \mu[\beta,\alpha] = 0.$$

Therefore $h(\alpha) \in (\pi^i K_{i+1}(A))^o = \widehat{K_i(A) + \pi A}$, as before.

By iterating $(7)_i$, we obtain:

$$h^j(K_\ell(A)) \subseteq \widehat{K_{\ell-j}(A) + \pi A}$$

and, therefore $h^d(A) \subseteq \widehat{\pi A}$. Iterating again, we have $h^{dj}(A) \subseteq \widehat{\pi^j A}$, and so $h^{d^2}(A) = 0$. In these arguments, we use the obvious inclusion $h(\widehat{B}) \subseteq \widehat{h(B)}$, for any submodule B and homomorphism h.

To prove $(5)_n$ and $(6)_i$ we made the following computation:

$$\begin{aligned}
<\Psi\alpha,\Psi\beta>_A &= <\alpha,\beta>_A + <h\alpha,\beta>_A + <\alpha,h\beta>_A + <h\alpha,h\beta>_A \\
&= <\alpha,\beta>_A + \epsilon\overline{<\beta,h\alpha>}_A + <\alpha,h\beta>_A + \epsilon\overline{<h\beta,h\alpha>}_A \\
&= <\alpha,\beta>_A + \epsilon\overline{\{\beta,\alpha\}} + \{\alpha,\beta\} + \epsilon\overline{\{h\beta,\alpha\}} \\
&= <\alpha,\beta>_A + [\alpha,\beta] + \epsilon\overline{\{h\beta,\alpha\}} \\
&= <\Phi\alpha,\Phi\beta>_B + \epsilon\overline{\{h\beta,\alpha\}}
\end{aligned}$$

For $(5)_n$ it remains to prove $\{\gamma,\alpha\} = 0$ for $\alpha \in \pi^{n-1}K_n(A)$, $\gamma \in K_n(A) + \pi A$. If $\gamma \in \pi A$, then $\{\gamma,\alpha\} = \{\gamma',\pi\alpha\} = 0$, since $\pi\alpha = 0$. If $\gamma \in K_n(A)$, then $\{\gamma,\alpha\} = \mu[\gamma,\alpha] = 0$ by $(1)_{n-1}$. This completes the proof of $(5)_n$.

To prove $(6)_i$, we must show $\{h\beta,\alpha\} = 0$ for $\alpha \in \pi^i K_{i+1}(A)$, $\beta \in K_{i+1}(A)$. But, by $(7)_i$, $h\beta \in \widehat{K_i(A) + \pi A}$, so it suffices to prove $\{\gamma,\alpha\} = 0$ if $\alpha \in \pi^i K_{i+1}(A)$, $\gamma \in K_i(A) + \pi A$. Writing $\alpha = \pi^i\alpha'$, we have $\{\gamma,\alpha\} = \{\pi^i\gamma,\alpha'\} = 0$ if $\gamma \in K_i(A)$ and $\{\gamma,\alpha\} = \{\pi\gamma',\alpha\} = \{\gamma',\pi\alpha\} = 0$, if $\gamma = \pi\gamma'$, since $\alpha \in K_1(A)$.

Now define $\Phi' = \Phi \circ \Psi^{-1}$. Then $(5)_n$ implies $(2)_{n-1}$ immediately, and $(6)_i$ implies $(1)_i$, where Φ is now replaced by Φ' in $(2)_{n-1}$ and $(1)_i$.

After such successive modifications of Φ we can assume Φ satisfies $(1)_i$, for all i, and $(2)_o$ which we rewrite as the case n = 1 of: $(8)_n$ $\langle\Phi\alpha,\Phi\beta\rangle_B = \langle\alpha,\beta\rangle_A$ for $\alpha \in k_n(A)$, $\beta \in A$. Note that $(8)_n$, for any $n \geq 1$, implies $(1)_i$, for all i since $\pi^i K_{i+1}(A) \subseteq K_1(A) \subseteq K_n(A)$.

We assume Φ satisfies $(8)_n$, for some $n \geq 1$, and modify Φ to satisfy $(8)_{n+1}$. When n=d, Φ will be the desired isometry.

We proceed as before. Define $[\alpha,\beta] = \langle\Phi\alpha,\Phi\beta\rangle_B - \langle\alpha,\beta\rangle_A$. By $(8)_n$, [,] is an ϵ-form on $A/K_n(A)$. Set $\{\alpha,\beta\} = \mu[\alpha,\beta]$; (3) is again satisfied. Since $K_n(A)^o = \widehat{\pi^n A}$, we can define a homomorphism $h:A \to \widehat{\pi^n A}$ such that:

$$\{\beta,\alpha\} = \langle\beta,h(\alpha)\rangle_A \text{ for all } \alpha,\beta \in A.$$

Clearly $h^j(A) \subseteq \widehat{\pi^{nj}(A)}$ and so h is nilpotent. Therefore $\Psi:A \to A$, defined by $\Psi(\alpha) = \alpha + h(\alpha)$ is an isomorphism. To check $(8)_{n+1}$, we compute, as above:

$$\langle\Psi\alpha,\Psi\beta\rangle_A = \langle\Phi\alpha,\Phi\beta\rangle_B + \{h\alpha,\beta\}$$

Since $\{\alpha,\beta\} = 0$ if $\beta \in K_n(A)$, it follows that $\{\alpha,\pi\beta\} = 0$ if $\beta \in K_{n+1}(A)$. Since $h(A) \subseteq \widehat{\pi^n A}$, we only need show $\{\alpha,\beta\} = 0$ if $\alpha \in \pi^n A$, $\beta \in K_{n+1}(A)$. If $\alpha = \pi^n\alpha'$, $n \geq 1$, then:

$$\{\alpha,\beta\} = \{\pi^n\alpha',\beta\} = \{\pi^{n-1}\alpha',\pi\beta\} = 0$$

This proves $(8)_{n+1}$, and, thereby, the theorem.

§21. Realization of product structure on homogeneous modules

In the case of A homogeneous of degree d, we have $\tilde{\Delta}_i(A) = 0$ if $i \neq d-1$, by (13.1), and $\tilde{\Delta}_{d-1}(A) = \Delta_{d-1}(A) = A_{d-1} = K_{d-1}(A)$. Thus there is only one non-trivial $<,>_i$ - for $i=d-1$ - which is defined on A_{d-1}. We have a sequence of inclusions:

$$A_{d-1} \subseteq A_{d-2} \subseteq \dots \subseteq A_0$$

of R-torsion free R modules of the same rank. Therefore, we may consider them as R-submodules of the Q(R)-vector space

$$V = A_{d-1} \otimes_R Q(R) = \dots = A_0 \otimes_R Q(R)$$

The R-valued ϵ-form $<,>_{d-1}$ on A_{d-1}, extends uniquely to a non-singular Q(R) valued ϵ-form [,] on V. The following two theorems give a complete characterization of the induced form [,].

Theorem (21.1): Let A be a homogeneous π-primary Λ-module of degree d with an $S(\Lambda)$ valued ϵ-form $<,>$, and suppose $\pi = \bar{\pi}$ is workable. If [,] denotes the induced ϵ-form on the Q(R)-vector space $V = A_{d-1} \otimes_R Q(R)$, then A_i is the R-dual of A_{d-1-i}, for each i, if and only if $<,>$ is non-singular on A.

Theorem (21.2): Let $A(d-1) \subseteq A(d-2) \subseteq \dots \subseteq A(1) \subseteq A(0)$ be R-torsion free of the same rank and [,] a non-singular ϵ-form on $V = A(d-1) \otimes_R Q(R)$ such that A(i) is the R-dual of A(d-1-i), for all i. If $\pi = \bar{\pi}$ is Dedekind, and R has a half-unit, then there exists a homogeneous π-primary Λ-module A of degree d with non-singular ϵ-form <,> such that $A_i = A(i)$, for all i and <,> induces [,].

By theorem (20.1), <,> is uniquely determined by [,], up to an automorphism of A. Note that we do not assert that A can be specified in advance.

The R-dual of a submodule B of V is $\{\alpha \in V\colon [\alpha,B] \subseteq R\}$.

<u>Proof of Theorem (21.1):</u> By definition, $[\pi^{d-1}r,\pi^{n-1}t] = \pi^{d-1}\langle r,t\rangle$, for any $r,t \in A$, where we have identified $A_{d-1} \cong \pi^{d-1}A$. The inclusion $A_{d-1} \subseteq A_i$ can be identified with the inclusion $\pi^{d-1}A \subseteq \pi^i K_{i+1} \approx A_i$. The pairing $[,]_i : A_i \times A_{d-1-i} \to R$ defined by $[\pi^i r,\pi^{d-1-i}t]_i = \langle r,t\rangle$ is obviously an extension of [,] on $A_{d-1} \times A_{d-1}$, and therefore agrees with [,] by the uniqueness of the extension. Note that:

$$[\alpha,\beta]_i = \epsilon\overline{[\beta,\alpha]}_{d-1-i}$$

We conclude that $A_{d-1-i} \subseteq$ R-dual of A_i. Suppose < , > is non-singular, and let $\alpha \in$ R-dual of A_i. Then $[\ ,\alpha]$ defines a homomorphism $A_i \to R^{1/\pi} \subseteq S(\Lambda)$. Identifying $A_i \approx \pi^i K_{i+1}$, we obtain a homomorphism $\phi\colon K_{i+1} \to S(\Lambda)$ by $\phi(\beta) = [\pi^i\beta,\alpha]$. By (19.2), ϕ extends over A; by non-singularity of < , >, there exists $\alpha' \in A$ such that $\phi = \langle\ ,\alpha'\rangle$ on K_{i+1}. Since $\phi(K_i) = 0$, we have $\alpha' \in (K_i)^0 = \widehat{\pi^i A} = K_{d-i}$, by homogeneity. Now

$$[\pi^i\beta,\pi^{d-1-i}\alpha'] = \langle\beta,\alpha'\rangle = \phi(\beta) = [\pi^i\beta,\alpha], \text{ for } \beta \in K_{i+1}.$$

Since [,] is non-singular and $\pi^i K_{i+1} = A_i$ is a full lattice in V, we conclude $\alpha = \pi^{d-1-i}\alpha' \in \pi^{d-1-i} K_{d-i} = A_{d-1-i}$.

Suppose A_i is the R-dual of A_{d-1-i}, for all i; ie. $[\ ,\ ]_i$ is non-singular for all i, in the sense that $[\ ,\ ]_i$ is adjoint to an isomorphism $\bar{A}_{d-1-i} \approx \mathrm{Hom}_R(A_i, R)$. Then $<\ ,\ >$ is certainly non-degenerate, since $<\alpha, A> = 0$ implies $<\pi^i\alpha, A> = 0$ and, if $\pi^{i+1}\alpha = 0$, this implies $[\pi^i\alpha, A_{d-1-i}]_i = <A, K_{d-i}(A)> = 0$. By non - degeneracy of $[\ ,\ ]_i$, we conclude $\pi^i\alpha = 0$.

To prove non-singularity of $<\ ,\ >$ consider a homomorphism $\Phi: A \to S(\Lambda)$. Suppose, inductively, that there exists $\alpha \in A$ such that $<.,\alpha> = \Phi$ on $K_i(A)$. If $\beta \in K_{d-i}(A)$, then $<K_i(A), \beta> = 0$, since $K_i(A)^o = \pi_i(A) = K_{d-i}(A)$. We will find β such that $<.,\beta> = \phi - <.,\alpha>$ on $K_{i+1}(A)$; then $<.,\alpha+\beta> = \phi$ on $K_{i+1}(A)$, completing the inductive step.

Clearly $\phi - <.,\alpha> = 0$ on $K_i(A)$ and so defines a homomorphism $K_{i+1}(A)/K_i(A) = A_i \to R$. Since $[\ ,\ ]_i$ is non-singular, there exists $\beta' \in A_{d-1-i}$ such that $[.,\beta']_i = \phi - <.,\alpha>$ on $K_{i+1}(A)$. If $\beta \in K_{d-i}(A)$ satisfies $\beta \equiv \beta'$ mod $K_{1-i-1}(A)$, then:

$<.,\beta> = [.,\beta']_i = \phi - <.,\alpha>$ on $K_{i+1}(A)$, as desired.

This completes the proof of (21.1).

<u>Proof of (21.2):</u> Let W be a free Γ_d-module of the same rank as

V, where $\Gamma_d = \Lambda_\pi/(\pi^d)$. We will identify V with $\pi^{d-1}W$. Let $<\ ,\ >$ be an ϵ-form on W with values in Γ_d, related to $[\ ,\ ]$ by the formula:

$$[\pi^{d-1}\alpha, \pi^{d-1}\beta] = \pi^{d-1}<\alpha, \beta> \quad \text{for } \alpha, \beta \in W$$

where we identify $\pi^{d-1}\Gamma_d$ with Q(R). The existence of such $<\ ,\ >$ is established by lifting a matrix representative M of $[\ ,\ ]$ to a matrix $\tilde{M}$ over Γ_d such that $\pi^{d-1}\tilde{M} = M$ and M is ϵ-Hermitian. The only problem in this is lifting the diagonal elements, i.e., given $\lambda \in Q(R)$ with $\lambda = \epsilon\bar{\lambda}$, does there exist $\tilde{\lambda} \in \Gamma_d$ such that $\pi^{d-1}\tilde{\lambda} = \lambda$ and $\tilde{\lambda} = \epsilon\bar{\tilde{\lambda}}$. But for this choose any $\tilde{\lambda}$ with $\pi^{d-1}\lambda = \lambda$ and then consider $\mu\tilde{\lambda} + \bar{\mu}\bar{\tilde{\lambda}}$ where $\mu + \bar{\mu} = 1$.

We conclude from the nonsingularity of $[\ ,\ ]$ that $<\ ,\ >$ is also nonsingular. In fact, $\det \tilde{M} \equiv \det M \bmod \pi\Gamma_d$. Since M is nonsingular, this means $\det \tilde{M}$ is a unit in Γ_d -- recall Λ_π is a discrete valuation ring.

If A is a finitely generated Λ_d-module in W, where $\Lambda_d = \Lambda/(\pi^d) \subseteq \Gamma_d$, such that $\Gamma_d A = W$, we will say A is <u>full in W</u>. The Λ_d-dual of $A = \{\alpha \in W\colon <\alpha, A> \subseteq \Lambda_d\}$ will also be a full Λ_d-module in W. If A is contained in its Λ_d-dual, then $<\ ,\ >$ induces a Λ_d-valued ϵ-form on A, which is clearly nondegenerate. Furthermore this induced form is nonsingular if and only if A and its Λ_d-dual are equal--the proof is immediate, since $A \otimes_{S_d} \Gamma_d = \Gamma_d A = W$.

Since any ϵ-form on a Λ-module of degree d takes values in $\Lambda_d \subseteq Q(\Lambda)/\Lambda$ (inclusion defined by dividing by π^d), the proof of (21.2) amounts to finding a full Λ_d-module A in W such that

$A_i = \pi^i A \cap V = A(i)$, for every i, and such that A is its own Λ_d-dual under < , >.

<u>Lemma (21.3)</u>: Let B be a homogeneous π-primary Λ-module of degree d and let $A_{d-1} \subsetneq A_{d-2} \subseteq \cdots \subseteq A_o \subseteq B_o$ be any sequence of submodules of $B_o = \pi^{d-1}B$. Then there exists a submodule A of B such that A_i is the i-th lower derivative of A ie. $\pi^i K_{i+1}(A) = A_i$.

<u>Proof</u>: Suppose, inductively, that we have $C \subseteq K_k(B)$ (k<d) such that $\pi^i K_{i+1}(C) = A_i$ for every i<k. Let $P \subseteq C$ be an elementary module of degree k such that $\pi^{k-1}P = A_k$, and $\tilde{P}$ any elementary module of degree k+1 with $\pi\tilde{P} \approx P$. Then the isomorphism $\pi\tilde{P} \approx P$ can be extended to an imbedding of $P \_ B$ by using (15.2) to complete the diagram:

$$\begin{array}{ccc} \tilde{P} & \dashrightarrow & L \\ \downarrow\pi & & \downarrow\pi \\ \pi\tilde{P} & \approx & P \end{array}$$

where $L = \{\alpha \in B: \pi\alpha \in P\}$; since B is homogeneous of degree > k, $\pi L = P$.

Now consider $C' = C + \tilde{P}$. Since $K_k(\tilde{P}) \subsetneq C$, we have $K_k(C') = C$ so $\pi^i K_{i+1}(C') = \pi^i K_{i+1}(C) = A_i$ for i<k. $\pi^k C' = \pi^k \tilde{P} = \pi^{k-1}P = A_k$. Thus C' satisfies the inductive requirements and the proof is complete.

<u>Lemma (21.4)</u>: Let W be a free Γ_d-module and $W^* = \mathrm{Hom}_{\Gamma_d}(W, \Gamma_d)$. Let $A \subseteq W$ be a finitely-generated Λ_d-module, and $A^* \subseteq W^*$ the Λ_d-dual

of A,ie. $A^* = \{\phi: \phi(A) \subseteq \Lambda_d\}$. Then $A^{**} = A$, where

$$A^{**} = \{\alpha \in W: \phi(\alpha) \in \Lambda_d \text{ for all } \phi \in A^*\}.$$

Proof: If $\gamma \notin A$, it suffices to find a homomorphism $\phi: W \to \Gamma_d$ such that $\phi(A) \subseteq \Lambda_d$ and $\phi(\gamma) \notin \Lambda_d$. We proceed by induction on d.

If $\gamma \in \pi W$, then by induction there exists $\phi': \pi W \to \Gamma_{d-1}$ with $\phi'(A \cap \pi W) \subseteq \Lambda_{d-1}$ and $\phi'(\gamma) \notin \Lambda_{d-1}$. By lemma (19.2), $\phi'|A \cap \pi W$ extends to $\tilde{\phi}: A \to \Lambda_d$, where $\Lambda_{d-1} = \pi\Lambda_d$ and $\Lambda_d \subseteq S(\Lambda)$ via division by π^d. Note that any homomorphism $A \to S(\Lambda)$ has image $\subseteq \Lambda_d$. Now $\tilde{\phi}$ and ϕ' extend to a homomorphism $\phi: W \to \Gamma_d$. First $\tilde{\phi}$ extends uniquely to $\phi'': \Gamma_d A \to \Gamma_d$; since ϕ'' and ϕ' agree on $\Gamma_d A \cap \pi W$, we thus have a homomorphism $\pi W + \Gamma_d A \to \Gamma_d$. But any homomorphism Ψ from a submodule B of a Γ_d-module W into Γ_d extends over W by the following argument. Since Γ_d is a principal ideal domain with ideals (π^i), there exists a direct sum decomposition $W = \bigoplus W_{i,j}$ for which $B = \bigoplus \pi^i W_{i,j}$, where $W_{i,j}$ is a free Γ_j-module, $j \leq d$. If $\{\alpha_\eta\}$ is a basis of $W_{i,j}$, then $\pi^{j-i}\Psi(\alpha_\eta) = 0$. Thus we may write $\Psi(\alpha_\eta) = \pi^i\lambda_\eta$ and the homomorphism $\alpha_\eta \to \lambda_\eta$ defines the desired extension.

Suppose $\gamma \notin \pi W$. If $\pi^{d-1}\gamma \in \pi^{d-1}A$, then, for some $\alpha \in A$, $\gamma - \alpha \in \pi W$. But now we can apply the preceding paragraph, using $\gamma - \alpha$ instead of γ. The resulting ϕ also has the property

$$\phi(\gamma) \notin \Lambda_d.$$

If $\pi^{d-1}\gamma \notin \pi^{d-1}A$, it suffices to find $\phi': \pi^{d-1}W \to Q(R)$ with $\phi'(\pi^{d-1}A) \subseteq R$, $\phi'(\pi^{d-1}\gamma) \in R$ and then define $\phi = \phi' \circ \pi^{d-1}$. Thus we are reduced to the case $d = 1$, where $\Gamma_d = Q(R) = F$, $\Lambda_d = R$.

Let $A' = A + R\alpha$. If $FA \neq FA'$, then there is an obvious non-zero ϕ: $FA' \to F$ such that $\phi(FA) = 0$, which will do. If $FA = FA' = W'$, then, by the theory of Dedekind domains there is a basis of W', $\alpha_1,\ldots,\alpha_n$, fractional ideals $J_1,\ldots,J_n,I_1,\ldots I_n$ such that $I_i \subsetneq J_i$ so that $A' = \Sigma J_i\alpha_i$ and $A = \Sigma I_i\alpha_i$ (See [ ]). Since $A \neq A'$ some $I_j \neq J_j$. Choose $\lambda \in F$ such that $\lambda I_j \subseteq R$ but $\lambda J_j \subsetneq R$. Then ϕ can be defined by $\phi(\alpha_i) = 0 \quad i \neq j$, $\phi(\alpha_j) = \lambda$. Clearly $\phi(A) \subseteq R$ and $\phi(A') \subsetneq R$; therefore $\phi(\alpha) \notin R$.

We now return to the proof of (21.2). Let k be the integer such that $d = 2k+1$ or $2k$. Then $[A(k),A(k)] \subseteq R$ and $A(k)$ is its own R-dual if $d-1 = 2k$.

Claim: There exists a full Λ_d-module $B \subseteq W$ such that:

(i) B is elementary of degree d with $B_o = \ldots = B_{d-1} = A(k)$

(ii) $\langle B,B\rangle \subseteq \Lambda_d$

The existence of B satisfying (i) is clear, since such B exists abstractly and the isomorphism $\pi^{d-1}B \approx A(k)$ can be extended to an imbedding by projectivity of B and $A(k) \subseteq \pi^{d-1}W$.

We will alter B by automorphisms of W to achieve (ii). Suppose inductively that $\langle B,\pi^iB\rangle \subseteq \Lambda_d$. This is true for $i = d-1$ already, since $\langle B,\pi^{d-1}B\rangle = [B_{d-1},B_{d-1}] \subseteq R$. Therefore assume $i \leq d-1$. Define $h: B \to Hom_\Lambda(B,\Gamma_d/\Lambda_d)$ by $h(\beta)\cdot\alpha = \mu\langle\alpha,\beta\rangle \bmod \Lambda_d$, where $\mu + \bar{\mu} = 1 \bmod \pi^d$.

By inductive assumption we can consider h as a map:

$$h: B/\pi^iB \to Hom_\Lambda(B/\pi^iB,\Gamma_i/\Lambda_i)$$

Since B is elementary of degree d, B/π^iB is elementary of degree i and so projective over Λ_i. Thus the natural map:

$$\mathrm{Hom}_\Lambda(B/\pi^iB,\Gamma_i) \to \mathrm{Hom}_\Lambda(B/\pi^iB,\Gamma_i/\Lambda_i)$$

is surjective. By projectivity of B/π^iB again, h lifts to a map:

$$\tilde{h}: B/\pi^iB \to \mathrm{Hom}_\Lambda(B/\pi^iB,\Gamma_i) \cong \mathrm{Hom}_\Gamma(W/\pi^iW,\Gamma_i)$$

The last isomorphism follows from $B/\pi^iB \otimes_{S_i} \Gamma_i = W/\pi^iW$. Since $<\ ,\ >$ is non-singular $\tilde{h}$ is adjoint to a mapping:

$$\tilde{\phi}: B/\pi^iB \to W$$

where $<\alpha,\tilde{\phi}(\beta)> = \tilde{h}(\beta)\cdot\alpha$, for any $\alpha \in W$, $\beta \in B$ identifying $\Gamma_i = \pi^{d-i}\Gamma_d$. Clearly image $\tilde{\phi} \subseteq \pi^{d-i}W$. Since $i \leq d-1$, $\tilde{\phi}$ is nilpotent and so $\Phi = 1 - \tilde{\phi}$ is an isomorphism of W. Let $B' = \Phi(B)$; $\pi^{d-1}B' = \pi^{d-1}B$ so (i) still holds.

We will show that $<B',\pi^{i-1}B'> \subseteq \Lambda_d$, completing the inductive step. Let $\alpha,\beta \in B$:

$$
\begin{aligned}
<\Phi(\alpha),\Phi(\beta)> &= <\alpha,\beta> - <\tilde{\phi}(\alpha),\beta> - <\alpha,\tilde{\phi}(\beta)> - <\tilde{\phi}(\alpha),\tilde{\phi}(\beta)> \\
&\equiv <\alpha,\beta> - \epsilon\overline{\mu}\overline{<\beta,\alpha>} - \mu\overline{<\alpha,\beta>} - <\tilde{\phi}(\alpha),\tilde{\phi}(\beta)> \mod \Lambda_d \\
&= -<\tilde{\phi}(\alpha),\tilde{\phi}(\beta)> \mod \Lambda_d
\end{aligned}
$$

by ϵ-Hermitian property of $<\ ,\ >$. But image $\tilde{\phi} \subseteq \pi^{d-i}W$, and so $<\tilde{\phi}(\alpha),\tilde{\phi}(\beta)> \subseteq \pi^{2d-2i}\Gamma_d$. Therefore, if $\alpha \in \pi^{i-1}B$, then $<\tilde{\phi}(\alpha),\tilde{\phi}(\beta)> \subseteq \pi^{2d-2i+i-1}\Gamma_d$, and, since $2d-i-1 \geq d$ if $i \leq d-1$, we conclude $<\tilde{\phi}(\alpha),\tilde{\phi}(\beta)> = 0$ -- ie. $<\Phi(\alpha),\Phi(\beta)> \in \Lambda_d$.

This proves the claim.

We now apply Lemma (21.3) to find a submodule $C \subseteq B$ such that

$$C_i = \begin{cases} A(i) & i \geq k \\ A(k) & i \leq k \end{cases}$$

Let C' be the Λ_d-dual of $C + \pi^{d-k}W$; so $C' \subseteq \pi^k W$. Note that $C' \subseteq C + \pi^{d-k}W$, and $<\pi^k C', C> \subseteq \Lambda_d$ and so $\pi^k C' = 0$ if $d = 2k$, or $\pi^k C' = C'_k \subseteq$ R-dual of $C_{d-k-1} = C_k = A(k)$ if $d = 2k + 1$. Since the R-dual of $A(k)$ is $A(k+1) \subseteq A(k) = C_k$, we have $\pi^k C' \subseteq \pi^k C$. Thus $C' \subseteq C + \pi^{d-k}W$ as desired.

One consequence of this is that $<C',C'> \subseteq \Lambda_d$, since $<C',C> \subseteq \Lambda_d$ and $<\pi^{d-k}W, \pi^{d-k}W> \subseteq \pi^{2d-2k}\Gamma_d = 0$ $(2d-2k \geq d)$. Now define $A = C + C'$; $<A,A> \subseteq \Lambda_d$ clearly.

Suppose $\alpha \in \Lambda_d$-dual of A. Thus $\alpha \in \Lambda_d$-dual of $C' = C + \pi^{d-k}W$; this follows from (21.4) by noticing that $\pi^k W$ is dually paired with $W/\pi^{d-k}W$ by $< \, , \, >$ and $C' = (C/C \cap \pi^{d-k}W)^*$. Let $\gamma \in C$ so that $\alpha - \gamma \in \pi^{d-k}W$. Since $\alpha \in \Lambda_d$-dual of C and $\gamma \in C \subseteq \Lambda_d$-dual of C, we have $\alpha - \gamma \in \Lambda_d$-dual of C. Furthermore $<\pi^{d-k}W, \pi^{d-k}W> \subseteq \pi^{2d-2k}W = 0$, since $2d-2k \geq d$. Thus $\alpha - \gamma \in \Lambda_d$-dual of $\pi^{d-k}W$ also, and therefore $\alpha - \gamma \in C'$. This shows $\alpha \in C + C' = A$, which implies A is its own Λ_d-dual.

We now show that $A_i = A(i)$, for all i. If $i \geq k$, then $\pi^i A = \pi^i C$, since $\pi^i C' \subseteq \pi^i(C + \pi^{d-k}W) = \pi^i C$. Therefore $A_i = C_i = A(i)$. If $i < k$, then A_i is the R-dual of A_{d-1-i} under $[\, , \,]$ by Theorem (21.1). But $d-1-i \geq k$, for $i < k$, and so A_i is the R-dual of $A(d-1-i)$, and this is precisely $A(i)$ by assumption. This completes the proof of (21.2).

§22. Product structure on semi-homogeneous modules

The results of the previous section we now extend to characterize the collection $\tilde{\Delta}_i(A)$ with non-degenerate $< \ , \ >_i$ for every π-primary Λ-module which decomposes into a direct sum of homogeneous submodules. We will call such modules semi-homogeneous. As pointed out in (17.5), not every π-primary Λ-module is semi-homogeneous unless degree ≤ 2.

Theorem (22.1): If A is semi-homogeneous, then the homogeneous summands are uniquely determined by A.

Proof: Suppose $A = \bigoplus_{i=1}^{d} A(i) = \bigoplus_{i=1}^{d} B(i)$ are direct sum decompositions, where A(i), B(i) are homogeneous of degree i. We show $A(i) \simeq B(i)$, for every i, by induction on d.

Consider the homomorphism $\phi : A(d) \to B(d)$ defined by inclusion in A followed by projection; we show ϕ is an isomorphism. Consider

$$L_i(A) = \frac{\widehat{\pi^{d-i-1}A}}{K_i(A) \cap \widehat{\pi^{d-i-1}(A)}}$$

; if A is homogeneous of degree k, then

$$L_i(A) = \begin{cases} A_i & k = d \\ 0 & k < d \end{cases}$$

. Thus the inclusion $A(d) \subseteq A$ induces isomorphisms $A(d)_i = L_i(A(d)) \simeq L_i(A)$ and the projection $A \to B(d)$ induces isomorphisms $L_i(A) \approx L_i(B(d)) = B(d)_i$. Therefore the map $\phi : A(d) \to B(d)$ induces isomorphisms $A(d)_i \xrightarrow{\simeq} B(d)_i$ for all i. By successive applications of the 5-lemma, this implies ϕ is an isomorphism.

Now consider $\Psi : B' \to A'$ -- where $B' = \bigoplus_{i=1}^{d-1} B(i)$, $A' = \bigoplus_{i=1}^{d-1} A(i)$ -- defined by inclusion in A followed by projection on A'. It follows from ϕ being an isomorphism that Ψ is also an isomorphism. If $\Psi(\beta') = 0$ then $\beta' \in A(d)$, and then $\phi(\beta') = 0$ which implies $\beta' = 0$. If

$\alpha' \in A'$, write $\alpha' = \beta' + \beta_o$ for $\beta' \in B'$, $\beta_o \in B(d)$; since ϕ is onto $\beta_o + \beta'' \in A(d)$ for some $\beta'' \in B'$. Thus $\alpha' - (\beta_o + \beta'') = \beta' - \beta''$, which implies $\alpha' = \Psi(\beta' - \beta'')$.

Since $B' \approx A'$, the inductive assumption implies $A(i) \approx B(i)$ for every $i < d$. The proof is now complete.

Theorem (22.2): If A is a semi-homogeneous π-primary module with non-singular ϵ-pairing $< , >$, then A decomposes into an orthogonal direct sum of homogeneous submodules, which are uniquely determined up to isometry.

Proof: Let $A = B \oplus C$, where B is homogeneous of degree d and C of degree $< d$. We will show $< , >|B$ is non-singular. Then, by standard arguments, $A = B \oplus B^o$; so $B^o \approx C$ and the theorem follows by induction.

We first show $< , >|B$ is non-degenerate. Let $\alpha \in B$ and suppose $<\pi^i\alpha, B> = 0$ for some $i < d$. Then $<\pi^{d-1}\alpha, A> = 0$ and so $\pi^{d-1}\alpha = 0$. Since $K_{d-1}(B) = \widehat{\pi B}$, we have $\lambda\alpha = \pi\alpha'$, for some $\alpha' \in B$, $\lambda \in \Lambda$ not divisible by π; clearly $<\pi^{i+1}\alpha', B> = 0$ since λ is prime to π. By induction we find $\alpha \in \pi^{d-i}B$; the case $i = 0$ implies non-degeneracy of $< , >|B$.

Suppose $\phi : B \to Q(\Lambda)/\Lambda$; we find $\beta \in B$ such that $\phi(\alpha) = <\alpha, \beta>$ for all $\alpha \in B$. Assume $\pi^i\phi = 0$ and proceed by induction on i. The case $i = 0$ is clear. For the general case, extend ϕ over A by setting $\phi(C) = 0$; then $\pi^i\phi = 0$ still. Since $< , >$ is non-singular on A, choose $\alpha \in A$ such that $\phi = <\cdot, \alpha>$. If we write $\alpha = \beta + \gamma$, $\beta \in B$, $\gamma \in C$ - then we prove $\gamma \in K_{i-1}(C)$. The condition $\pi^i\phi = 0$ means

$\phi(\pi^i B) = 0$ and so $\phi(\widehat{\pi^i B}) = 0$. But $\widehat{\pi^i B} = K_{d-i}(B)$ and so there exists Ψ: $\pi^{d-i}B \to Q(\Lambda)/\Lambda$ such that $\Psi \circ \pi^{d-i} = \phi$. By $\lambda\Psi$ extends to a homomorphism Ψ': $B \to Q(\Lambda)/\Lambda$, for some $\lambda \in \Lambda$ not divisible by π; then Ψ' extends over A by $\Psi'(C) = 0$. It is clear that $\lambda\phi = \Psi' \circ \pi^{d-i}$. If $\alpha' \in A$ satisfies $\Psi' = \langle .,\alpha' \rangle$, then $\pi^{d-i}\alpha' = \lambda\alpha$. If $\alpha' = \beta' + \gamma'$ - $\beta' \in B, \gamma' \in C$ - then $\lambda\gamma = \pi^{d-i}\gamma'$, and so $\gamma \in \widehat{\pi^{d-i}C} \subseteq K_{i-1}(C)$, as claimed, since $\pi^{d-1}C = 0$. Now consider the homomorphism $\phi' : B \to Q(\Lambda)/\Lambda$ defined by $\langle .,\gamma \rangle$; $\gamma \in K_{i-1}(C)$ implies $\pi^{i-1}\phi' = 0$ and so, by induction there exists $\theta \in B$ such that $\langle .,\gamma \rangle = \langle .,\theta \rangle$ on B. Now we see that,

on B: $\phi = \langle .,\beta \rangle + \langle .,\gamma \rangle = \langle .,\beta \rangle + \langle .,\theta \rangle = \langle .,\beta+\theta \rangle$.

To complete the proof of (22.2) we must prove that if $A = \bigoplus_i A(i) = \bigoplus_i B(i)$ are two orthogonal decompositions into homogeneous summands, then A(i) and B(i) are isometric for every i.

We show that the isomorphisms ϕ and Ψ of Theorem (22.1) can be changed to isometries by (20.1). Recall that, for a homogeneous module C of degree k, $\tilde{\Delta}_i(C) = 0$ if $i \neq k - 1$, thus $\tilde{\Delta}_i(A) = \tilde{\Delta}_i(A(i+1)) = \tilde{\Delta}_i(B(i+1))$. So $\tilde{\Delta}_i(A') = \tilde{\Delta}_i(A) = \tilde{\Delta}_i(B')$ for $i < d - 1$, and $\tilde{\Delta}_{d-1}(A(d)) = \tilde{\Delta}_{d-1}(A) = \tilde{\Delta}_{d-1}(B(d))$; these equalities are obviously induced by inclusions or projections. Thus ϕ and Ψ induce these identifications, which are obviously isometries of the ϵ-pairings $\langle\ ,\ \rangle_i$ induced by $\langle\ ,\ \rangle$. Theorem (20.1) then allows us to replace ϕ and Ψ by isometries of the ϵ-pairings induced on A', B', A(d), B(d).

The proof of (22.3) is now completed by induction on d.

§23. A non-semi-homogeneous module

As a consequence of these results and (17.4) we now have a

complete classification of π-primary Λ-modules of degree ≤ 2 with non-singular ϵ-pairing, if π is Dedekind and R has a half-unit - at least up to classical questions of integral Hermitian from theory (see [J]). For degree 3 we know by (17.5) that not every module is semi-homogeneous and, therefore, the characterization given by (21.1) and (21.2) is unavailable, although the faithfulness of the invariants is true by (17.2) and (20.1). For degree ≥ 4, we have seen, by (18.1), that the π-primary sequences are insufficient for classification.

Of course, there is the possibility that the examples (17.5) and (18.1) do not extend to the category of π-primary Λ-modules with non-singular ϵ-pairings. To remedy this we now provide such examples.

Given any finitely-generated π-primary Λ-module A of degree d with π-only torsion, let $A^* = \mathrm{Hom}_\Lambda(A,\Lambda_d)$. If we set $B = A \oplus A^*$, there is an obvious ϵ-pairing defined on B by the formula:

$$\langle\alpha,\beta\rangle = \begin{cases} 0 & \text{if } \alpha,\beta \in A \text{ or } \alpha,\beta \in A^* \\ \beta(\alpha)\pi^{-d} & \text{if } \alpha \in A,\ \beta \in A^* \\ \epsilon\overline{\alpha(\beta)}\,\overline{\pi}^{-d} & \text{if } \alpha \in A^*,\ \beta \in A \end{cases}$$

Since A has π-only torsion, $A \subseteq A \otimes_{\Lambda_d} \Gamma_d$ which can be imbedded in a free Γ_d-module, since Γ_d is a discrete valuation ring, we can apply (21.4) to conclude $\langle\ ,\ \rangle$ is non-singular.

<u>Proposition (23.1)</u>: If A is the Λ-module of (17.5), then B is not semi-homogeneous.

<u>Proof</u>: Let V represent the free Γ_3-module of rank 2; we may regard

A on the Λ_3-module in V generated by $(\bar{\sigma}^2,0),(0,\pi\bar{\sigma})$ and (π^2,π^2). $V^* = \mathrm{Hom}_\Gamma(V,\Gamma_3)$ is also free of rank 2 and we may represent V^* as pairs $[\lambda,\mu]$ acting on V by the formula:

$$[\lambda,\mu]\cdot(\theta,t) = \lambda\theta + \mu t$$

It is easy to check $A^* \subseteq V^*$ is generated by the elements $[1,0]$, $[\pi\bar{\sigma}^{-2},0]$ and $[-\bar{\sigma}^{-1},\pi\bar{\sigma}^{-1}]$ as Λ_3-module.

Consider $B_2 \subseteq B_1 \subseteq B_0 = B \cap \pi^2 W \subseteq \pi^2 W$ where $W = V \oplus V^*$. Identifying $\pi^2\Gamma_3 = Q(R)$, we may check that:

B_0 is generated by $(\sigma,0)$, $(1,1)$, $[\sigma^{-2},0]$ and $[0,\sigma^{-1}]$

B_1 is generated by $(\sigma^2,0)$, $(0,\sigma)$, $[\sigma^{-2},0]$ and $[0,1]$

B_2 is generated by $(\sigma^2,0)$ and $[\sigma^{-1},0]$

as R-modules.

Suppose $B = B(1) \oplus B(2) \oplus B(3)$, where $B(i)$ is homogeneous of degree i. Obviously $B_2 = B(3)_2$ and so we may compute $B(3)_i$ by the formula $B(3)_i = \widehat{B(3)_2} \cap B_i$. We find $B(3)_1$ is generated by $(\sigma^2,0)$ and $[\sigma^{-2},0]$ and $B(3)_0$ by $(\sigma,0)$ and $[\sigma^{-2},0]$.

Notice also that $B_0 \subseteq \widehat{B_1}$. This implies $B(1) = 0$, since $B_0 = B(1)_0 \oplus B(2)_0 \oplus B(3)_0$ and $B_1 = B(2)_1 \oplus B(3)_1$.

Now we have $B_0 = B(2)_0 \oplus B(3)_0$; therefore, $B(2)_0$ must be generated by two elements:

$$\alpha = (1,1) + a(\sigma,0) + b[\sigma^{-2},0]$$

$$\beta = [0,\sigma^{-1}] + c(\sigma,0) + d[\sigma^{-2},0]$$

where $a,b,c,d \in R$. Notice that

$$\sigma\alpha = (\sigma,\sigma) + a(\sigma^2,0) + b\sigma[\sigma^{-2},0]$$

$$\sigma\beta = [0,1] + c(\sigma^2,0) + d\sigma[\sigma^{-2},0]$$

are elements of B, and, therefore, of $B(2)_1$. We will show that they, in fact, must generate $B(2)_1$ by showing that the order ideal of $B(2)_o/B(2)_1$ is (σ^2).

One computes easily that the order ideal of B_o/B_1 is (σ^3) while that of $B(3)_o/B(3)_1$ is (σ). Since $B_o/B_1 = B(2)_o/B(2)_1 \oplus B(3)_o/B(3)_1$ the desired result follows.

Now consider the element $(0,\sigma) \in B_1$. We show this is <u>not</u> in $B(2)_1 + B(3)_1$. Suppose $(0,\sigma) = \lambda\sigma\alpha + \mu\sigma\beta + \gamma(\sigma^2,0) + \eta[\sigma^{-2},0]$. By considering only the components in V we have:

$$(0,\sigma) = \lambda((\sigma,\sigma) + a(\sigma^2,0)) + \mu c(\sigma^2,0) + \gamma(\sigma^2,0)$$

The second coordinate of this formula implies $\lambda=1$. Now the 1st coordinate gives the formula:

$$0 = \sigma + a\sigma^2 + \mu c\sigma^2 + \gamma\sigma^2$$

or, on dividing by σ,

$$0 = 1 + \sigma(a + \mu c + \gamma)$$

For such an equation to hold it is necessary that σ be a unit in R, which it is not.

This proves (23.1).

<u>Theorem (23.2)</u>: If I, I' are the ideals of Λ_4 as defined in (18.1) and $B = I \oplus I^*$, $B' = I' \oplus (I')^*$, then B and B' are not isomorphic as Λ-modules, but their π-primary sequences are isomorphic.

<u>Proof</u>: Recall that I and I' are the Λ-modules in Γ_4 generated by $\sigma^4, \sigma^3\pi, \sigma\pi^2$ and π^3 (for I) and $\sigma^4 + \sigma^2\pi, \sigma^3\pi, \sigma\pi^2$ and π^3 (for I'). Identifying I^* and $(I')^*$ with submodules of Γ_4 as above, we check that I^* is generated by $1, \sigma^{-1}\pi, \sigma^{-3}\pi^2$ and $\sigma^{-4}\pi^3$ and $(I')^*$ is generated by $1, \sigma^{-1}\pi, \sigma^{-3}\pi^2 - \sigma^{-5}\pi^3$ and $\sigma^{-4}\pi^3$.

Note that $I^* \approx I$ -- in fact, $\sigma^4 I^* = I$.

We now compute the Fitting invariants of B and B' and show they are different. We leave it to the reader to confirm that presentation matrices of I,I' and $(I')^*$ are given as follows:

$$I: \begin{pmatrix} \pi & -\sigma & 0 & 0 \\ 0 & \pi & -\sigma^2 & 0 \\ 0 & 0 & \pi & -\sigma \\ 0 & 0 & 0 & \pi \end{pmatrix}$$

$$I': \begin{pmatrix} \pi & -\sigma & -\sigma & 0 \\ 0 & \pi & -\sigma^2 & 0 \\ 0 & 0 & \pi & -\sigma \\ 0 & 0 & 0 & \pi \end{pmatrix}$$

$$(I')^*: \begin{pmatrix} \pi & -\sigma & 0 & 0 \\ 0 & \pi & -\sigma^2 & -\sigma \\ 0 & 0 & \pi & -\sigma \\ 0 & 0 & 0 & \pi \end{pmatrix}$$

These use the generators in the order given above and a presentation matrix (λ_{ij}) corresponds to a complete set of relations $\sum_j \lambda_{ij}\alpha_j = 0$, where $\{\alpha_j\}$ are the generators.

Computing the minors of the above matrices gives the results:

$$E_0(I) = E_0(I') = E_0((I')^*) = 0$$

$$E_1(I) = I$$

$$E_1(I') = E_1((I')^*) = I'$$

Using the general formula $E_k(M \oplus N) = \sum_{i=0}^{k} E_i(M)E_{k-i}(N)$ we conclude

$E_1(B) = I^2$, $E_1(B') = (I')^2$. It suffices, therefore, to show $I^2 \neq I'^2$.

Now I^2 is generated by the elements: $\sigma^8, \sigma^7\pi, \sigma^5\pi^2$ and $\sigma^4\pi^3$. On the other hand $(I')^2$ contains the element $(\sigma^4 + \sigma^2\pi)\sigma\pi^2 = \sigma^5\pi^2 + \sigma^3\pi^3$. If $I^2 = (I')^2$, then I^2 contains $(\sigma^5\pi^2 + \sigma^3\pi^3) - \sigma^5\pi^2 = \sigma^3\pi^3$; but this contradicts the given generators of I^2. Thus $B \not\cong B'$.

Finally we prove easily that the π-primary sequences of B and B' are isomorphic. By (18.1), those of I and I' are isomorphic and, therefore, those of I^* and $(I')^*$ are also isomorphic. Since π-primary sequences behave in the obvious way under connected sum, our assertion is proved.

This proves (23.2).

§24. Rational classification of product structure

Although we do not obtain a complete characterization of the ϵ-forms $\langle\ ,\ \rangle_i$, except in the semi-homogeneous case, we will be able to obtain such a characterization for the "rational" invariants-ie. the extensions of $\langle\ ,\ \rangle_i$ to ϵ-forms over Q(R), the quotient field of R.

If A is a π-primary Λ-module with π-only torsion and $\langle\ ,\ \rangle$ an ϵ-form on A, we may consider the extension of $\langle\ ,\ \rangle$ to an ϵ-form, still denoted $\langle\ ,\ \rangle$, on $A \otimes_\Lambda \Gamma = W$, where Γ is the localization of Λ at π, with values in $Q(\Lambda)/\Gamma$. Equivalently, if A has degree d, we may regard $\langle\ ,\ \rangle$ as taking values in $\Lambda_d \subseteq Q(\Lambda)/\Lambda$ and the extension on W as taking values in $\Gamma_d \subseteq Q(\Lambda)/\Gamma$. The classification of ϵ-forms over Γ_d has a classical reduction to forms over Q(R) --

(see [M1]). We quote that result here in a slightly different notation.

If W is a finitely-generated π-primary Γ-module, we can consider the π-primary sequences as usual. In this case, however, things are simpler and it is not hard to see that the $\{\Delta_i(W)\}$ contain all the information. If W supports an ϵ-form $\langle\ ,\ \rangle$, we have as usual the induced ϵ-forms $\langle\ ,\ \rangle_i$ on $\Delta_i(W) = \tilde{\Delta}_i(W)$.

It is easy to see $\langle\ ,\ \rangle$ is non-degenerate (which is the same as non-singular here) if and only if every $\langle\ ,\ \rangle_i$ is non-degenerate.

<u>Theorem (24.1)</u>: If $(W, \langle\ ,\ \rangle)$ and $(W', \langle\ ,\ \rangle')$ are two non-singular ϵ-forms on π-primary Γ-modules, then they are isometric if and only if each $\Delta_i(W)$ is isometric to $\Delta_i(W')$.

The proof of this theorem is in [M1], and we leave to the reader the exercise of showing that our formulation is equivalent to Theorem of [M1].

On the other hand, given ϵ-forms $\langle\ ,\ \rangle_i$ on $\Delta_i(W)$ it is easy to see, from the arguments in [M1], how to construct an ϵ-form $\langle\ ,\ \rangle$ on W inducing the given $\langle\ ,\ \rangle_i$.

Returning to our π-primary Λ-module A with ϵ-form $\langle\ ,\ \rangle$, with $W = A \otimes_\Lambda \Gamma$, it is easy to see that $\Delta_i(W) \approx \Delta_i(A) \otimes_R Q(R)$ and the $\langle\ ,\ \rangle_i$ on $\Delta_i(W)$ induced from the extension of $\langle\ ,\ \rangle$ over W is the same as the extension over $\Delta_i(W)$ of the ϵ-form $\langle\ ,\ \rangle_i$ on $\Delta_i(A)$ induced from $\langle\ ,\ \rangle$. Thus the "rational" invariants of $(A, \langle\ ,\ \rangle)$ consists of the isometry classes of the extensions over $\Delta_i(W)$ of the $\langle\ ,\ \rangle_i$ on $\Delta_i(A)$.

The question to ask now is the following. Which collections of ϵ-forms $\langle\ ,\ \rangle_i$ on vector spaces V_i over $Q(R)$ arise from a non-singular ϵ-form on a π-primary Λ-module.

Theorem (24.2): Let V_i be a $Q(R)$-vector space with non-singular ϵ-form $\langle\ ,\ \rangle_i$, for $i=1,\dots,d$. If $\pi = \bar{\pi}$ is Dedekind, and R contains a half-unit, then there exists a π-primary Λ-module A with non-singular ϵ-form $\langle\ ,\ \rangle$ such that $\Delta_i(W)$ is isometric to V_i (and 0 for $i > d$) if and only if $\bigoplus_i V_{2i}$ contains a self-dual lattice L (ie. L is a finitely-generated R-module in $\bigoplus_i V_{2i}$ which is its own R-dual under $\bigoplus_i \langle\ ,\ \rangle_{2i}$.

Note that (24.2) follows easily from (21.1) and (21.2) for the homogeneous case. In fact A homogeneous of degree $(k+1)$ corresponds to $V_i = 0$ for $i \neq k$. If k is odd, there is no restriction on the form $\langle\ ,\ \rangle_k$ while, for k even, V_k must contain the self-dual lattice $A_{k/2}$.

Proof: Suppose A is of degree d with non-singular $\langle\ ,\ \rangle$. We show $\bigoplus_i \Delta_{2i}(W)$ contains a self-dual lattice by induction on d. If d=1, then $\bigoplus_i V_{2i} = W$ and $L = A$ makes the theorem obvious.

If $d > 1$, consider the Λ-module $B = \dfrac{K_{d-1}(A)}{\widehat{\pi^{d-1}A}}$ which is of degree $< d$. By (19.5) and (19.6), the ϵ-form induced on B by $\langle\ ,\ \rangle$ is non-singular, if $\langle\ ,\ \rangle$ is non-singular. Clearly $X = B \otimes_\Lambda \Gamma = \dfrac{K_{d-1}(W)}{\pi^{d-1}W}$. A straight forward computation gives $\Delta_i(X) = 0$ for $i \geq d-1$, and $\Delta_i(X) \approx \Delta_i(W)$ for $i < d-3$ and $i = d-2$, isometric with respect to the induced ϵ-pairings. Also we have $\Delta_{d-3}(X) = \dfrac{K_{d-2}(W) + \pi W}{K_{d-3}(W) + \pi k_{d-1}(W)}$. Obviously $\Delta_{d-3}(X)$ contains $\Delta_{d-3}(W) = K_{d-2}(W)/(K_{d-3}(W) + \pi K_{d-1}(W))$ as a subspace with quotient $\pi W/K_{d-2}(W) \cap \pi W \approx \Delta_{d-1}(W)$. Furthermore this decomposition is consistent with the induced pairings and so, by the Gram-Schmidt process, we obtain an isometry:

$$\Delta_{d-3}(X) \approx \Delta_{d-3}(W) \oplus \Delta_{d-1}(W)$$

As a consequence of this diagnosis we have an isometry:

$$\bigoplus_i \Delta_{2i}(X) \approx \bigoplus_i \Delta_{2i}(W)$$

If $<\ ,\ >$ on A - and so, also, on B - is non-singular, we conclude, by induction, that $\bigoplus_i \Delta_{2i}(X)$ contains a self-dual lattice. Thus we complete the inductive step of this half of the Theorem.

For the converse, suppose given $V_i, <\ ,\ >_i$ as in the statement of the Theorem. Set $W = \oplus W_i$, where W_i is a free Γ_i-module of rank = dimension V_i; thus $W_i/\pi W_i \approx V_i$. Let $[\ ,\ ]_i$ be a non-singular ϵ-form on W_i inducing $<\ ,\ >_i$; and define $<\ ,\ >$ on W as the orthogonal sum of $[\ ,\ ]_i$. We must find $A \subset W$, finitely-generated Λ-module which is <u>full</u> in W - ie. $\Gamma A = W$ -- such that $<\ ,\ >|A$ is <u>integral</u> -- ie. $<A,A> \subseteq \Lambda_d \subseteq \Gamma_d$ -- and non-singular. This latter property can be checked by showing $A = A^*$, where $A^* \subseteq W$ is the set of all α such that $<\alpha,A> \subseteq \Lambda_d$, since $<\ ,\ >$ is non-singular on W and every homomorphism $A \to \Lambda_d$ extends to $W \to \Gamma_d$.

Let $V \subseteq W$ be the submodule defined by $V = \bigoplus_i \pi^i W_{2i+1}$. The pairing on $V/\pi V$ induced by $\langle\ ,\ \rangle$ is clearly isometric to $\bigoplus_i \Delta_{2i}(W)$ and, therefore, by assumption contains a self-dual lattice L. Choose a finitely-generated Λ-module $C \subseteq V$ so that $C \equiv L \bmod \pi V$; clearly $\langle C,C\rangle = \langle L,L\rangle \subseteq \Lambda_d$.

Let D' be any finitely-generated Λ-module full in W. We may choose $\lambda \in \Lambda$ not divisible by π so that $D = \lambda D'$ satisfies:

(i) $\langle D,D\rangle \subseteq \Lambda_d$

(ii) $D \cap V \subseteq C$

(iii) $\langle D,C\rangle \subseteq \Lambda_d$

This follows from fullness of C and finite-generated property of C and D. Now set $B = C + D$; B is full in W and finitely-generated, satisfying:

(i)' $\langle B,B\rangle \subseteq \Lambda_d$

(ii)' $B \cap V = C$

The pairing $\langle\ ,\ \rangle$ of $U = \bigoplus \pi^i W_{2i}$ induces a non-singular pairing: $(U \oplus \pi V) \times (W/U \oplus V) \to \Gamma_d$. Let B' be the Λ_d-dual in $U \oplus \pi V$ of B mod $U \oplus V$, and set $A = B + B'$. Since $\langle B',B'\rangle \subseteq \langle \pi V, \pi V\rangle = 0$, we have $\langle A,A\rangle \subseteq \Lambda_d$ -- ie. $A \subseteq A^*$. We show $A^* \subseteq A$.

Note that $A \cap (U \oplus \pi V) = B'$, since (i)' implies $B \cap \pi V \subseteq B'$. If $\alpha \in A^*$, then $\langle \alpha, B'\rangle \subseteq \Lambda_d$ and so $\alpha \in B \bmod U \oplus V$ by (21.4). Therefore, since $B \subseteq A$ we may assume $\alpha \in U \oplus V$. Since $\langle \alpha, C\rangle \subseteq \Lambda_d$ and L is self-dual, we conclude $\alpha \in L \equiv C \bmod U \oplus \pi V$. Thus we may assume $\alpha \in U \oplus \pi V$. But now $\langle \alpha, B\rangle \subseteq \Lambda_d$ implies $\alpha \in B' \subseteq A$.

The proof of (24.2) is now complete.

Our main interest is in the case $\Lambda = Z[t,t^{-1}]$, in which case

Q(R) is a global field and R is, therefore, the ring of integers corresponding to some Dedekind set of primes on Q(R). We will now give an alternative formulation of (24.2) covering this case.

Recall the classification of Hermitian forms over a global field E (see e.g. [La]). Let F be the fixed field and E_P, F_p denote the completions of E, F at primes P, p.

(i) r = rank, a positive integer

(ii) $d_p = \pm 1$, defined for every discrete prime p of F which extends to a single prime P of E.

(iii) $\sigma_p \in Z$, defined for every real spot p on F whose extensions over E are complex.

r is the dimension of the underlying vector space; $d_p = \pm 1$ if and only if the determinant is a local norm at p (see [O]); σ_p is the signature of the complex Hermitian form on $E_P = \mathbb{C}$. It follows from the Hasse norm theorem that the determinant, as an element of $\dot{F}/N_{E/F}(\dot{E})$, is determined by $\{d_p, \sigma_p\}$.

These invariants are subject to the following relations:

(a) $|\sigma_p| \leq r$

(b) $\sigma_p \equiv r \mod 2$

(c) $\prod_p d_p = \prod_p (-1)^{(r-\sigma_p)/2}$

Conversely any collection $\{r, d_p, \sigma_p\}$ obeying these relations arises from some Hermitian form.

A classification of skew-Hermitian forms follows from the observation that scaling a skew-Hermitian form by a non-zero $\mu \in \dot{E}$, satisfying $\bar{\mu} = -\mu$, converts it into a Hermitian form. By taking the determinant and signature of this associated Hermitian form we obtain a complete set of invariants independent of μ which we denote by the same symbols.

If F is the fixed field of the conjugation, then μ is well defined up to multiplication by an element of $\dot{F}$. Clearly $E = F(\mu)$, where $\mu^2 = \theta \in F$. For each prime p of F, consider the exponent of p in the prime factorization of the principal ideal generated by θ -- define $o_p(E/F) = +1$ if this is even, -1 if odd. If E is unramified at p, then $o_p(E/F) = +1$; if E is ramified at a non-dyadic p, then $o_p(E/F) = -1$ (see [J:§5]).

Lemma 2.4.3: Given an ϵ-form over $E = Q(R)$, a global field, with R a Dedekind domain, then there exists a self-dual lattice (over R) if and only if:

i) $d_p = +1$ for every discrete prime p of $R \cap F$ at which E is unramified and d_p is defined,

ii) r is even, if $\epsilon = -1$ and $o_p(E/F) = -1$ for any discrete prime p of $R \cap F$,

iii) $d_p = (\theta, -1)_p^s$, if $\epsilon = -1$, for every discrete non-dyadic prime p of $R \cap F$ at which E is ramified and d_p is defined, where $r = 2s$ (see (ii)).

Proof: If p is a discrete prime of F which extends to a single prime P of E, then $P = \overline{P}$. Therefore, the conjugation on E extends to one on E_p with fixed field F_p, and induces one on the residue field K_p. An ϵ-form over E will, then, extend to one over E_p. If L is an R-lattice, and P is a prime of R, then L/PL inherits an induced ϵ-form, which is defined over K_p. If L is self-dual at P, then the induced ϵ-form over K_p is non-singular, and conversely.

<u>Claim</u>: An ϵ-form admits a self-dual lattice over R, if and only if its extension over E_P admits a self-dual lattice, for every discrete self-conjugate prime P of R.

<u>Proof of Claim</u>: If L is self-dual over R, then L_P is self-dual at P. Conversely suppose there are self-dual lattices at all P as described. Let M be any full R-lattice; then $M \subseteq M^{\#}$, the R-dual of M. We may choose a vector space basis $\{\alpha_i\}$ so that $M^{\#} = \sum_i I_i J_i \alpha_i$, where I_i are fractional ideals of R, and J_i are integral ideals. We may write $J_i = T_i S_i \overline{S}_i$, where every prime dividing T_i is self-conjugate. By replacing M with $\sum_i S_i^{-1}\alpha_i$, we may assume that $S_i = R$, for all i -- ie. $I_i = T_i$.

For every prime P of R dividing any I_i, we assume there exists a lattice L_P at E_P which is self-dual. Now choose L so that $L = L_P$ at these primes P and L = M at all other primes ([0:81:14]). Thus $L^{\#} = L$ (see [0:§82F,K]), as desired.

In view of this result we may assume E is a local field and p the only prime of F. If there exists a self-dual lattice L, we may use a basis of L to compute the determinant Δ -- thus Δ will be a unit of E. If E/F is unramified, then every unit of F is a norm [0:(63:16)]; thus $d_p = +1$, if $\epsilon = +1$ or if $\epsilon = -1$ and r is even. Since $o_p(E/F) = +1$, we can choose μ to be a unit and so $\mu\Delta$ is also a unit; thus $d_p = +1$ in general, when E/F is unramified.

Suppose E/F is ramified. Then there exist units of F which are not norms; for if not then, for any element $\pi \in E$ generating the ideal P, $\pi\overline{\pi}$ is a norm generating P and so every element of F would be a norm. By [0:63.13a] this is not true.

If $o_p(E/F) = -1$, then we may choose μ to be a generator of P; thus $\mu\Delta$ will also generate P. But if $\epsilon = -1$, and r is odd, $\mu\Delta \in F$ and, since E/F is ramified, must generate an even power of P. Thus, since $\mu\Delta$ defined d_p if r is odd, we conclude r is even.

The induced conjugation on the residue field K is trivial, if E/F is ramified (see [J: §5]). Thus, if $\epsilon = -1$ and L is a self-dual lattice, the induced form on L/PL is skew-symmetric and non-singular over K. Therefore its determinant, which coincides with Δ mod P, is square in K. If p is nondyadic, then $o_p(E/F) = -1$ and so r is even. Therefore $\Delta \in F$ and is square mod p. By Hensel's lemma, since p is nondyadic, Δ is a square of F. Thus Δ is certainly a norm, and so the determinant of the (+1)-form obtained by scaling μ is $\mu^r\Delta = \theta^s\Delta$. Therefore $d_p = (\theta^s\Delta, \theta)_p = (\theta^s, \theta)_p = (\theta, \theta)_p^s = (\theta, -1)_p^s$.

For the converse, first note that the diagonal matrices $(1,1,\ldots,1)$, for $\epsilon = +1$, and $(\mu,\ldots,\mu)$, for $\epsilon = -1$, define ϵ-forms of any rank r with $d_p = +1$. If E/F is ramified, there exists a unit λ of F which is not a norm and the diagonal matrices, $(\lambda,1,\ldots,1)$, for $= +1$, and $(\lambda\mu,\mu,\ldots,\mu)$, for $\epsilon = -1$, define ϵ-forms of any rank r with $d_p = -1$. In the cases of $\epsilon = -1$, we can choose μ a unit $o_p(E/F) = +1$ and so the lattice generated by the obvious basis elements in all these examples is self-dual.

If E/F is ramified and p nondyadic, a sum of s hyperbolic planes, i.e., (-1)-forms with representative matrix $\begin{pmatrix} 0 & 1 \\ -1 & 0 \end{pmatrix}$ defines a (-1)-form with $r = 2s$ and $d_p = (\theta, -1)_p^s$. If p is dyadic and $o_p(E/F) = -1$, we may choose μ to generate P. Since K has characteristic

2, every element is a square and so any unit α of F can be written in the form: $\alpha = x^2 + y\theta$, where x is a unit, and y an integer of F. Thus α is the determinant of the (-1)-form defined by $\begin{pmatrix} \mu & x \\ -x & y\mu \end{pmatrix}$.

If α is chosen not to be a norm, then a sum of this form and hyperbolic planes defines a (-1)-form with $d_p = -(\theta,-1)^s_p$ and an arbitrary even rank r.

This completes the proof of (24.3).

We now return to the situation where we have a π-primary Λ-module A supporting a non-degenerate ϵ-form < , >. If Q(R) is a global field, then according to (24.1) the Γ-module $A \otimes_\Lambda \Gamma = W$, with the extended ϵ-form, is classified by the following invariants:

(i) $r_i = \text{dimension}_E\ \Delta_i(W)$ $(E = Q(R))$

(ii) $d_{p,i} = \pm 1$ defined from $\Delta_i(W)$, for every discrete prime p of F (fixed field) which extends to a single prime P of E.

(iii) $\sigma_{p,i} \in Z$ defined from $\Delta_i(W)$ for every real spot p on F, whose extensions over E are complex.

We have the relations:

(a) $|\sigma_{p,i}| \leq r_i$

(b) $\sigma_{p,i} \equiv r_i \mod 2$

(c) $\prod_p d_{p,i} = \prod_p (-1)^{(r_i - \sigma_{p,i})/2}$

As a direct consequence of (24.2) and (24.3), we have:

<u>Corollary (24.4)</u>: If Λ is a UFD with involution and $\pi = \bar{\pi}$ a Dedekind prime of Λ such that $R = \Lambda/(\pi)$ contains a half-unit and Q(R) is a global field, then a set of elements $\{r_i, d_{p,i}, \sigma_{p,i}\}$

arise as invariants, defined above, from a π-primary Λ-module A with non-singular ϵ-form if and only if they satisfy (a)-(c) above and:

(d) $\prod_i d_{p,2i} = +1$ for every prime p of $R \cap F$ at which E is unramified.

(e) $\sum_i r_{2i}$ is even, for $\epsilon = -1$, if $o_p(E/F) = -1$ for any prime p of $R \cap F$.

(f) $\prod_i d_{p,2i} = +1$, if $\epsilon = -1$, for every non-dyadic prime p of $R \cap F$ at which E is ramified.

Note that, by (22.2), if A were semi-homogeneous then conditions (d), (e), (f) would have to be strengthened by changing " $\prod_i d_{p,2i}$ " to " each $d_{p,2i}$ " and " $\sum_i r_{2i}$ " to " each r_{2i} ". This observation permits one to give many examples of non-semi-homogeneous modules.

§25. Non-singular lattices over a Dedekind domain

Given an ϵ-form over $E = Q(R)$ a global field, lemma (24.3) gives us necessary and sufficient conditions for the existence of a self-dual lattice. The variability of this lattice is determined by the following:

Theorem (25.1): Let R be a Dedekind domain and A a finitely-generated torsion free R-module supporting an ϵ-form self-dually. Then we may choose $\Delta \in Q(R)$ to represent the determinant (extended over Q(R)) and I the ideal class of A, so that $I\bar{I} = \Delta R$ as fractional ideals of R.

Conversely, suppose we are given an ϵ-form over Q(R) with determinant Δ which has a self-dual lattice. If I is any fractional ideal of R such that $I\bar{I} = \Delta R$, then there exists a self-dual lattice whose ideal class is represented by I.

Remark: The restriction on Δ represented by the equation $(\Delta) = I\overline{I}$ corresponds to (i) of lemma (24.3), in case Q(R) is a global field.

Proof: Let $V = A \otimes Q(R)$ support the extended ϵ-form. We may choose a basis $\alpha_1,\ldots,\alpha_r$ of V such that $A \supseteq F$, the free R-module generated by $\{\alpha_i\}$. If $\Delta \in R$ is the determinant of the ϵ-form with respect to $\{\alpha_i\}$, then it is a standard fact that (Δ) is the order ideal of $F^{\#}/F$ ($F^{\#}$ is the dual of F). Thus, we can conclude from the inclusions $F \subseteq A = A^{\#} \subseteq F^{\#}$, the equality $(\Delta) = IJ$, where I is the order ideal of A/F and J the order ideal of $F^{\#}/A^{\#}$. But $F^{\#}/A^{\#} \approx \mathrm{Ext}^1_R(\overline{A}/\overline{F}, R) \simeq \overline{A/F}$ since R is a Dedekind domain and A/F is a finitely-generated torsion module. Thus $J = \overline{I}$. Finally we observe that, in general, the order ideal of a quotient A/B represents $\dfrac{\text{ideal class of A}}{\text{ideal class of B}}$ in the ideal class group of R -- since F is free, the proof of the first part of lemma (25.1) is complete.

Suppose L is a self-dual lattice of an ϵ-form over the quotient field Q(R). Then we have $I_o\overline{I}_o = (\Delta_o)$, where I_o represents the ideal class of L and Δ_o the determinant of the form. Given I, Δ as in the hypothesis we can write $\Delta = \Delta_o\lambda\overline{\lambda}$, for some $\lambda \in Q(R)$, and so $I\overline{I} = (\lambda I_o)(\overline{\lambda I_o})$. Thus we may rephrase the problem as follows. Given any ideal K such that $K^{-1} = \overline{K}$ $(K = I(\lambda I_o)^{-1})$, can we find another self-dual lattice L' such that:

(ideal class of L') = K·(ideal class of L)?

If we write $K = J_1 J_2^{-1}$, where J_i are integral ideals of R then $K^{-1} = \overline{K}$ is equivalent to $J_1 = \overline{J_2}$, ie. $K = J\overline{J}^{-1}$, for some integral ideal J. Without loss of generality, we may assume J is prime and

$J \neq \bar{J}$.

Suppose dim $V = 1$. Then we may write $L = I\alpha$, where α is a non-zero element of V, and L self-dual corresponds to the fact that $\langle\alpha,\alpha\rangle I\bar{I} = R$. In this case we may define $L' = KI\alpha$, and the proof is complete.

Now assume dim $V \geq 2$. Let $\alpha \in L$ generate a direct summand. Let $L_0 \subseteq L$ consist of all β such that $\langle\alpha,\beta\rangle \subseteq J$; then $\langle .,\alpha\rangle$ defines an isomorphism $L/L_0 \approx R/\bar{J}$. If $\langle\alpha,\alpha\rangle \in J\bar{J}$, we define $L' = (J^{-1}\alpha) + L_0$. In either case, we see easily that $\langle L',L'\rangle \subseteq R$. Furthermore we will prove that $L'/L_0 \approx R/J$. As a consequence we can then conclude that ideal class $L' = J \cdot$ (ideal class L_0); since ideal class $L = \bar{J}\cdot$(ideal class L_0), L' will have the desired ideal class.

Consider the case $\langle\alpha,\alpha\rangle \in J\bar{J}$. Since α generates a direct summand of L, it follows that $J^{-1}\alpha \cap L = R\alpha = J^{-1}\alpha \cap L_0$. Thus $L'/L_0 \approx J^{-1}\alpha/R\alpha \approx R/J$. For the case $\langle\alpha,\alpha\rangle \notin J\bar{J}$, we observe that $\bar{J}J^{-1}\alpha \cap L = R\alpha \cap \bar{J}J^{-1}\alpha = (R \cap \bar{J}J^{-1})\alpha = \bar{J}\alpha$, since $J \neq \bar{J}$ and both are prime. Since $\bar{J}\alpha \subset L_0$, we have $\bar{J}J^{-1}\alpha \cap L_0 = \bar{J}\alpha$ also. Thus $L'/L_0 \approx \bar{J}J^{-1}\alpha/\bar{J}J^{-1}\alpha \cap L_0 \approx \bar{J}J^{-1}\alpha/\bar{J}\alpha \approx R/J$.

It remains to check that L' is self-dual. In fact, it suffices to verify that the volume of L' is R (see [0:§82G]). But L is self-dual and so volume $L = R$; therefore the volume of L_0 is $J\bar{J}$, since $L/L_0 \approx R/\bar{J}$ [0:82.11]. But $L'/L_0 \approx R/J$ and so $J\bar{J}$ = volume L_0 = (volume L) $J\bar{J}$, which implies the desired result.

This completes the proof of Theorem (25.1).

Theorem (25.1) tells us exactly which isomorphism classes of R-modules can be realized as $A_{k/2}$ for A homogeneous of degree $k + 1$

(k even), when the isometry class of $W = A \otimes_{\Lambda} \Gamma$, or equivalently $\Delta_k(W)$, is prescribed.

§26. Norm criterion for a non-singular lattice

We give for the Hermitian or even-dimensional skew-Hermitian case, a useful reformulation of the relation $I\bar{I} = \Delta R$, of Theorem (25.1), between the ideal class of I-denoted [I] - and the class of $\Delta = \bar{\Delta}$ in the multiplicative groups $Q(R)/NQ(R)$, denoted $\{\Delta\}$; $N: \dot{Q}(R) \to \dot{Q}(R)$ is the norm defined by $N(\lambda) = \lambda\bar{\lambda}$.

Let R be a Dedekind ring with involution, and R_0 the subring of self-conjugate elements. Let $F \subsetneq E$ denote the quotient fields of $R_0 \subsetneq R$. If $I(R_0)$, $I(R)$ denote the ideal class groups of R_0, R, we have homomorphisms $S': I(R_0) \to I(R)$ and $N': I(R) \to I(R_0)$. S' is the extension induced by $I \mapsto IR$, for any ideal I of R_0, and N' is the norm induced by:

$$P \to \begin{cases} p^2 & \text{if } pR = P \\ p & \text{if } pR = P\bar{P} \end{cases}$$

for any prime ideal P of R. The standard relations are:

(i) $N' \circ S'(\alpha) = \alpha^2$, for any $\alpha \in I(R_0)$

(ii) $S' \circ N'(\alpha) = \alpha\bar{\alpha}$, for any $\alpha \in I(R)$

where $\alpha \mapsto \bar{\alpha}$ is the involution on I(R) induced by that on R.

We also have homomorphism $N: \dot{E} \to \dot{F}$ (the norm) and $S: \dot{F} \to \dot{E}$ (inclusion) and relations similar to (i), (ii) above. If $\dot{R}_0$, $\dot{R}$ denote the multiplicative groups of units in R_0, R, then we have inclusions $\dot{R}_0/N\dot{R} \subseteq \dot{F}/N\dot{E}$. We define a homomorphism:

$$\Phi: \text{ Kernel } N' \to \frac{\dot{F}/N\dot{E}}{\dot{R}_0/N\dot{R}} \quad .$$

If $N'[I] = 1$, we have $N(I) = \lambda R_0$ for some $\lambda \in \dot{F}$, and define $\Phi[I] = \lambda$ modulo $(N\dot{E})\dot{R}_0$. It is easy to see Φ is well-defined

Lemma 26.1: Let I be an ideal of R and $\lambda \in \dot{F}$. Then $I\bar{I} = \lambda' R$, for some $\lambda' \in \lambda \bmod N\dot{E}$ if and only if $N'[I] = 0$ and $\Phi[I] = \{\lambda\}$.

Proof: If $N'[I] = 0$, and $\Phi[I] = \lambda$ then $N(I) = \lambda' R_0$, as in the statement of the lemma. But then, by extending over R, we have $I\bar{I} = \lambda' R$.

Conversely, suppose $I\bar{I} = \lambda' R$. We must show $N(I) = \lambda' R_0$. But, if we consider the ideal $J = (1/\lambda')N(I)$ of R_0, then, by assumption, $JR = (1/\lambda')I\bar{I} = R$, which implies $J \equiv R_0$.

As an immediate consequence, we have the following reformulation of part of Theorem 25.1.

Theorem 26.2: Let V be a vector space over the field $E = Q(R)$, where R is a Dedekind domain with involution, with an ϵ-form --assume $\dim V$ is even if $\epsilon = -1$. Assume that V contains a self-dual lattice. Then, if $\Delta \in \dot{F}/N\dot{E}$ is the class of the determinant, where $F \subset E$ is the fixed field of the involution, and I is any ideal of R, V contains a self-dual lattice with ideal class $[I]$ if and only if $N'[I] = 0$ and $\Phi[I] = \{\Delta\}$.

We may derive a precise determination, from Theorem 26.2, of the number of different ideal classes of self-dual lattices with a given determinant from the following.

Proposition 26.3: Let $\psi: I(R) \to I(R)$ be the automorphism induced by conjugation of ideals of R. Then Image $(1 - \psi) \subseteq$ Kernel N' and Image $(1 - \psi)$ = Kernel Φ.

Proof: Since $N' \cdot \psi = N'$, i.e., $N'(\bar{\alpha}) = N'(\alpha)$ for any $\alpha \in I(R)$, the first inclusion follows immediately.

Suppose $\alpha \in I(R)$ is represented by the ideal I, and $N(I) = I_o$. Then $(1 - \psi)(\alpha)$ is represented by $I(\bar{I})^{-1}$ and $N(I(\bar{I})^{-1}) = I_o I_o^{-1} = (1)$. Thus $\Phi(1 - \psi)(\alpha) = 1$ and so Image $(1 - \psi) \subseteq$ Kernel Φ.

Suppose $\Phi(\alpha) = 0$, where α is represented by I. Write $I = \Pi P_i^{e_i} \Pi \bar{P}_i^{e'_i} \Pi Q_i^{f_i}$, the prime decomposition where $P_i \neq \bar{P}_i$, and $Q_i = \bar{Q}_i$. Then $N(I) = \Pi p_i^{e_i + e'_i} \Pi q_i^{f'_i}$ where P_i/p_i, Q_i/q_i and f'_i is f_i or $2f_i$ depending on whether q_i is unramified or ramified. Suppose $N(I)$ is principal generated by an element of $N\dot{E}$; we may then change I (within its ideal class) so that $N(I) = R_o$. In this case $e_i + e'_i = 0 = f'_i$ and so $e'_i = -e_i$, $f_i = 0$ and $I = \prod_i (P_i \bar{P}_i^{-1})^{e_i}$. Writing $J = \Pi P_i^{e_i}$, we have $I = J(\bar{J})^{-1}$ and so $\alpha \in$ Image $(1 - \psi)$.

Remark: The homomorphism Φ is just a generalization of the concept of genus [B-S: §8.4] and Proposition 26.3 the analogue of Theorem 7 of [B-S: p. 246].

§27. Dedekind criterion: p-adic reduction

In the next few sections, we will develop an effective criterion for $Z[\alpha, \alpha^{-1}]$ to be integrally closed, and, therefore, Dedekind. Z_p will denote the rational p-adic integers.

Lemma 27.1: $Z[\alpha, \alpha^{-1}]$ is integrally closed if and only if $Z_p[\alpha, \alpha^{-1}]$ is integrally closed, for every prime p.

This is a well-known fact.

Suppose $\phi(t)$ is a minimal polynomial for α, where not all coefficients are divisible by p. We may decompose $\phi(t) = \phi_1(t)\ldots\phi_k(t)$ over the p-adic integers $Z_{(p)}$, where $\{\phi_i(t)\}$ are relatively prime irreducible polynomials. Then $\phi_i(t) \equiv \Psi_i(t)^{e_i}$ mod p, for an irreducible polynomial $\Psi_i(t)$ in $Z/p[t]$. Let α_i be a root of $\phi_i(t)$ -- note that either α_i is a unit, in which case $Z_{(p)}[\alpha_i,\alpha_i^{-1}] = Z_{(p)}[\alpha_i]$, or $\Psi_i(t)$ is a monomial and $Z_{(p)}[\alpha_i,\alpha_i^{-1}] = Q_{(p)}[\alpha_i]$. This follows immediately from Hensel's lemma.

Lemma (27.2): If $Z_p[\alpha,\alpha^{-1}]$ is integrally closed, then $Z_{(p)}[\alpha_i,\alpha_i^{-1}]$ is integrally closed, for all i.

Since $Z_p[\alpha,\alpha^{-1}]$ contains the p-adic integers of $Q[\alpha]$, $Z_{(p)}[\alpha_i,\alpha_i^{-1}]$ contains the integers of $Q_{(p)}[\alpha_i]$. If α_i is a unit, this implies $Z_{(p)}[\alpha_i,\alpha_i^{-1}]$ is the ring of integers.

Lemma (27.3): If $Z_p[\alpha,\alpha^{-1}]$ is integrally closed, then, if $i \neq j$ and neither $\Psi_i(t)$ nor $\Psi_j(t)$ is a monomial, $\Psi_i(t)$ and $\Psi_j(t)$ are relatively prime.

Let D denote the p-adic integers of $Q[\alpha]$ and D_i the integers of $Q_{(p)}[\alpha_i]$. Then the natural map $D/pD \to \prod_i D_i/pD_i$ is an isomorphism (see e.g. [S]). If α_i is a unit (ie. $\Psi_i(t)$ not a monomial), then $D_i/pD_i \approx Z/p[t]/(\Psi_i(t)^{e_i})$. Since $D \subseteq Z_p[\alpha,\alpha^{-1}]$, there exists $f(t) \in Z_p[t,t^{-1}]$, such that $f(\alpha_i) \in pD_i$, $f(\alpha_j) \notin pD_j$. Thus $f(t) \in (\Psi_i(t)^{e_i})$, but $f(t) \notin (\Psi_j(t)^{e_j})$. If $e_i \leq e_j$, this implies $\Psi_i(t)$ and $\Psi_j(t)$ are relatively prime.

Theorem (27.4): $Z_p[\alpha,\alpha^{-1}]$ is integrally closed if and only if every $Z_{(p)}[\alpha_i,\alpha_i^{-1}]$ is integrally closed and the $\{\Psi_i(t)\}$ which are <u>not</u> monomials are pairwise relatively prime.

We have proved the "only if" part. Let $f(t) \in Q[t,t^{-1}]$ such that $f(\alpha)$ belongs to the integral closure of $Z_p[\alpha,\alpha^{-1}]$. By assumption, $f(\alpha_i) \in Z_{(p)}[\alpha_i,\alpha_i^{-1}]$, for each i. Therefore $f(t) \equiv f_i(t) \bmod \phi_i(t)$, for some $f_i(t) \in Z_{(p)}[t,t^{-1}]$. Now we may assume $f(t) = \frac{g(t)}{p^k}$ for some integer $k \geq 0$ and $g(t) \in Z_p[t,t^{-1}]$, and so $g(t) \equiv p^k f_i(t) \bmod \phi_i(t)$. By the Gauss lemma $g(t) - p^k f_i(t)$ is divisible by $\phi_i(t)$ in $Z_{(p)}[t,t^{-1}]$. If $k > 0$, we have $g(t)$ divisible by $\Psi_i^{e_i}(t)$ in $Z/p[t,t^{-1}]$. Since the $\Psi_i(t)$ are relatively prime, we may conclude that $g(t)$ is divisible by $\phi(t)$ in $Z/p[t,t^{-1}]$ and, therefore $g(t) = pf_1(t) + \phi(t)\lambda(t)$ for some $f_1(t)$, $\lambda(t) \in Z_p[t,t^{-1}]$. Now we see $f(\alpha) = \frac{g(\alpha)}{p^k} = \frac{f_1(\alpha)}{p^{k-1}}$. We may now replace $f(t)$ by $f_1(t)/p^{k-1}$ and repeat the above procedure k-1 more times to construct $f_i(t) \in Z_p[t,t^{-1}]$ satisfying:

$$f(\alpha) = \frac{f_1(\alpha)}{p^{k-1}} = \frac{f_2(\alpha)}{p^{k-2}} = \ldots\ldots = f_k(\alpha)$$

Therefore $f(\alpha) \in Z_p[\alpha,\alpha^{-1}]$.

§28. <u>A computable Dedekind criterion</u>

We next examine $Z_{(p)}[\alpha_i]$, when α_i is a unit. Let $\hat{\Psi}_i(t) \in Z_{(p)}[t]$ such that $\hat{\Psi}_i(t) \equiv \Psi_i(t) \bmod p$. Then $\phi_i(t) - \hat{\Psi}_i(t)^{e_i} = p\gamma_i(t)$, for some $\gamma_i(t) \in Z_{(p)}[t]$.

<u>Lemma (28.1)</u>: $Z_{(p)}[\alpha_i]$ is integrally closed if and only if $e_i = 1$ or $\gamma_i(\alpha_i)$ is a unit of $Q_{(p)}[\alpha_i]$.

$Z_{(p)}[\alpha_i]$ contains all the integers of $Q_{(p)}[\alpha_i]$ if and only if it contains a generator of the prime ideal P_i of $Q_{(p)}[\alpha_i]$, since the unit α_i generates the residue class field. Since e_i is the ramification index, we have $\hat{\Psi}_i(\alpha_i)^{e_i} = -p\gamma_i(\alpha_i) = P_i^{e_i}\gamma_i(\alpha_i)$. If $\gamma_i(\alpha_i)$ is a unit, then $\hat{\Psi}_i(\alpha_i)$ is the desired generator of P_i. If $e_i = 1$, then p is a generator of P_i.

Conversely, suppose $f(\alpha_i) = P_i$. Then f(t) is divisible by $\Psi_i(t)$, over Z/p, and so $f(t) = pg(t) + \hat{\Psi}_i(t)h(t)$ for some g(t), $h(t) \in Z_{(p)}[t]$. If $e_i > 1$, then $P_i = \hat{\Psi}_i(\alpha_i)h(\alpha_i)$ and so $\hat{\Psi}_i(\alpha_i) = P_i$ (since $\hat{\Psi}_i(\alpha_i) \in P_i$). But now the formula $\hat{\Psi}_i(\alpha_i)^{e_i} = P_i^{e_i}\gamma_i(\alpha_i)$ implies $\gamma_i(\alpha_i)$ is a unit.

We can now summarize the above results in the following effective procedure for determining whether $Z[\alpha,\alpha^{-1}]$ is Dedekind, where α is a root of the integral primitive polynomial $\phi(t)$, irreducible over Q.

For each prime p, write:

$$\phi(t) \equiv \epsilon t^k \prod_{i=1}^{n} \Psi_i(t)^{m_i} \quad \text{mod } p$$

where the mod p polynomials $\Psi_i(t)$ are pairwise relatively prime, irreducible, and not monomials and ϵ is a unit mod p. Choose an integral polynomial $\hat{\Psi}_i(t)$ such that $\hat{\Psi}_i(t) \equiv \Psi_i(t)$ mod p and write:

$$\phi(t) = \epsilon t^k \left(\prod_{i=1}^{n} \hat{\Psi}_i(t)^{m_i} \right) + p\,\gamma(t)$$

Theorem (28.2): $Z[\alpha,\alpha^{-1}]$ is Dedekind unless, for some prime p, there exists i such that $m_i > 1$ and $\gamma(t)$ is divisible by $\Psi_i(t)$ mod p.

Proof: Let $\phi(t) = \mu(t) \prod_{i=1}^{m} \phi_i(t)$ be a factorization, over the p-adic numbers, where $\mu(t)$ is a monomial mod p and each $\phi_i(t)$ is irreducible and not monomial mod p. By Hensel's lemma $\phi_i(t) \equiv \epsilon_i \Psi_i'(t)^{e_i}$ mod p, where $\Psi_i'(t)$ is irreducible and ϵ_i is a unit mod p. If $m > n$, then $\Psi_i'(t) = \Psi_j'(t)$, for some $i \neq j$, and by 27.1 and 27.4, $Z[\alpha, \alpha^{-1}]$ is not integrally closed. On the other hand, we may write $\phi_i(t) = \epsilon_i \Psi_i'(t)^{e_i} + p\gamma_i(t)$, and then $\gamma(t) \quad \sum_i (\gamma_i(t) \prod_{j \neq i} \epsilon_j \gamma_j'(t)^{e_j})$ mod p. If $\Psi_i'(t) = \Psi_j'(t)$, then it divides $\gamma(t)$ and, in addition, $e_i + e_j > 1$. Thus 28.2 holds when $m > n$.

If $m = n$, we may assume $\Psi_i'(t) = \Psi_i(t)$ and $e_i = m_i$. Writing $\phi_i(t) = \epsilon_i \hat{\Psi}_i(t)^{m_i} + p\gamma_i(t)$, we again see that $\gamma(t) \equiv \sum_i (\gamma_i(t) \prod_{j \neq i} \epsilon_j \Psi_j(t)^{m_j})$ mod p. If $\Psi_i(t) | \gamma(t)$ mod p, then $\Psi_i(t) | \gamma_i(t)$ mod p. By 27.4 and 28.1 the proof of 28.2 reduces to observing that, when $m_i > 1$, $\gamma_i(\alpha_i)$ is a unit of $Q_{(p)}[\alpha_i]$ if and only if $\Psi_i(t)$ does not divide $\gamma_i(t)$. But α_i generates the residue field. This, obviously, completes the proof.

We note, finally, that the criterion for $Z[\alpha, \alpha^{-1}]$ Dedekind, given by 28.2, is an effectively computable procedure. To see this, assume ϕ has been chosen to have no terms with negative exponent and nonzero constant term, Then ϕ is irreducible, in the usual sense, and Δ = discriminant ϕ is nonzero. It is only necessary now to check the conditions of 28.2 for primes which divide Δ. In fact, if the leading coefficient a_0 of ϕ is prime to p, then $\Delta \equiv$ discriminant ϕ_p mod p, while if a_0 is divisible by p, then $\Delta \equiv a_1^2$ discriminant ϕ_p, where a_1 is the next coefficient of ϕ. These formulae follow from the determinant definitions of discriminant (see [V: 5.8, 5.9]). In either case, if ϕ_p has a multiple factor, then Δ is divisible by p

§29. Computation of low-degree cases

To illustrate (28.2) we consider the low degree cases. Assume $\phi(t)$ is primitive and let Δ denote its discriminant.

1) Degree $\phi(t) = 2$: $Z[\alpha, \alpha^{-1}]$ is Dedekind unless, for some prime p, $p^2 | \Delta$, but at most one coefficient of $\phi(t)$ is divisible by p, and $\Delta \equiv 0$ or $4 \mod 16$ if $p = 2$.

In this case, the criterion of (28.2) reduces to the existence of a nontrivial double root a of $\phi(t) \pmod p$ such that $\phi(a)$ is divisible by p^2. It is not hard to see this is equivalent to the stated criterion.

2) Degree $\phi(t) = 3$: $Z[\alpha, \alpha^{-1}]$ is Dedekind unless for some prime p there exists an integer a such that

$$\phi'(a) \equiv 0 \mod p$$

$$\phi(a) \equiv 0 \mod p^2$$

(thus $p | \Delta$).

In order that $Z[\alpha, \alpha^{-1}]$ be not Dedekind, (28.2) implies we may write $\phi(t) = c(t - a)^2(t - b) + p\gamma(t)$ for some prime p, where $c, a \not\equiv 0 \mod p$, and $\gamma(a) \equiv 0 \mod p$. This is easily seen to be equivalent to the stated criterion.

3) Degree $\phi(t) = 4$; $\phi(t) = \phi(t^{-1})$: (This is satisfied when A supports a nondegenerate ϵ-pairing--see (19.1).) $Z[\alpha, \alpha^{-1}]$ is Dedekind unless $\phi(1)$ or $\phi(-1)$ is divisible by p^2, for some prime p, or, writing $\phi(t) = at^2 + bt + c + bt^{-1} + at^{-2}$, and, setting $\Delta = b^2 - 4a(c - 2a)$, $p^2 | \Delta$ for some prime p such that $p \nmid a$ and, if $p = 2$, $\Delta \equiv 0$ or $4 \mod 16$.

We omit the proof.

§30. Determination of ideal class group

The classification of ϕ-primary Λ-modules, when $R = \Lambda/\phi$ is Dedekind, requires computation of the ideal class group of $R = Z[\alpha, \alpha^{-1}]$, where α is a root of $\phi(t)$. In some cases this can be reduced to computations of the ideal class group of algebraic number fields.

Lemma 30.1: Suppose α is a root of the integral, irreducible polynomial $\phi(t)$. Then $Q \cap Z[\alpha, \alpha^{-1}]$ is generated (as a ring) by $\{\frac{1}{p}\}$, where p ranges over all primes for which $\phi(t)$ is a monomial mod p.

Proof: Let $S = Q \cap Z[\alpha, \alpha^{-1}]$; we first show $\frac{1}{p} \in S$, for primes p satisfying the stated condition. We have, for some integer a: $\phi(t) = at^m + pf(t)$, for some integral polynomial $f(t)$. Since $\phi(t)$ is primitive, a is not divisible by p, and, by a change in $f(t)$, we may assume $a = 1$. Thus:

$$0 = \phi(\alpha) = \alpha^m + pf(\alpha), \text{ and so } \frac{1}{p} = -\alpha^{-m} f(\alpha) \in Z[\alpha, \alpha^{-1}].$$

Conversely, suppose a/b is a rational number in $Z[\alpha, \alpha^{-1}]$, where a, b are relatively prime. Thus $a/b = f(\alpha)$, for some $f(t) \in Z[t, t^{-1}]$ or, equivalently;

$$a/b = f(t) + \phi(t)g(t), \text{ for some } g(t) \in Q[t, t^{-1}].$$

Then $a = bf(t) + b\phi(t)g(t)$; by Gauss' lemma, $h(t) = bg(t) \in Z[t, t^{-1}]$. If p divides b, then

$$a \equiv h(t)\phi(t) \mod p.$$

Regarding this as an equation over $Z/p[t, t^{-1}]$, and since a is relative prime to b, we conclude that $\phi(t)$ reduced mod p is a unit in $Z/p[t, t^{-1}]$. Thus it must be a monomial mod p, concluding the proof.

Let m be the product of all rational primes p such that $\phi(t)$ is a monomial mod p. Let M be the product of the first and last coefficients of $\phi(t)$. Note that $m|M$, since $\phi(t)$ is non-constant.

Proposition 30.2: If α, $\phi(t)$ as in (30.1), and $Z[\alpha, \alpha^{-1}]$ is integrally closed, then $\mathcal{O}[\frac{1}{m}] \subseteq Z[\alpha, \alpha^{-1}] \subseteq \mathcal{O}[\frac{1}{M}]$, where $\mathcal{O}$ is the ring of integers in $Q[\alpha]$.

Proof: The first inclusion follows from (30.1). The second follows from the observation that α and α^{-1} are integral over $Z[\frac{1}{M}]$.

For example, we may consider $\phi(t)$ satisfying:

(*) If p is a rational prime dividing the first or last coefficient of $\phi(t)$, then $\phi(t)$ is a monomial mod p.

This restriction has been considered by Crowell [C] in his study of the structure of Alexander modules. In this case m and M have the same prime divisors and so $\mathcal{O}[\frac{1}{m}] = \mathcal{O}[\frac{1}{M}]$. In fact, α and α^{-1} are integral over the ground ring $Z[\frac{1}{m}]$, and we have, without the Dedekind assumption:

Proposition 30.3: If $\phi(t)$ satisfies (*), then $Z[\alpha, \alpha^{-1}] = Z[\frac{1}{m}, \alpha] = Z[\frac{1}{m}, \alpha^{-1}]$.

Proof: That $Z[\alpha, \alpha^{-1}]$ contains the other two follows from (30.1).

Now, after multiplication by a power of t, we may write $\phi(t)$ in the form $\sum_{i=0}^{k} a_i t^i$, where $m = a_0 a_k$. Thus $\sum_{i=0}^{k} a_i \alpha^i = 0$, which can be rewritten $a_0 = -(\sum_{i=1}^{k} a_i \alpha^{i-1})$ or $a_k = -\alpha^{-1}(\sum_{i=0}^{k-1} a_i (\alpha^{-1})^{k-1-i})$. Since a_0 and a_k are units in $Z[\frac{1}{m}]$, we see that α is a unit in $Z[\frac{1}{m}, \alpha]$ and α^{-1} is a unit in $Z[\frac{1}{m}, \alpha^{-1}]$.

Proposition 30.3 provides an easier test of $Z[\alpha, \alpha^{-1}]$ Dedekind--for example, $Z[\alpha, \alpha^{-1}]$ will be Dedekind if the discriminant of $\phi(t)$ has no square factors prime to m. Another consequence of (30.2) or (30.3) is that the ideal class group of $Z[\alpha, \alpha^{-1}]$, when Dedekind and $\phi(t)$ satisfies (*), is the quotient of the ideal class group of $Q[\alpha]$ by the subgroup corresponding to those ideals which divide m.

§31. The quqdratic symetric case

We now work out, as an example of the preceding sections, the special case when ϕ is a quadratic "symmetric" polynomial: $\phi_a = at^2 + (1 - 2a)t + a$. This is the case of interest in knot theory-the symmetry is necessary for the existence of a nondegenerate ϵ-pairing (19.1).

To begin we observe that, by (30.3), $R_a = \Lambda/\phi_a$ is Dedekind if and only if the discriminant $1 - 4a$ is square-free. To classify R_a-modules, we must determine the "ideal class group of R_a, which is isomorphic to the quotient of the usual ideal class group of $F_a = Q(\sqrt{1 - 4a})$ by those classes represented by ideals dividing a. For the case $a > 0$, the order is effectively computable. Since the order of the ideal class group I of quadratic fields is computable (see [B]), we need only compute the order of the subgroup S defined by ideals dividing a. Suppose $a = p_1^{e_1} \cdots p_k^{e_k}$ is the prime decomposition; let P_i be a prime of F_a dividing p_i. Note that there is another prime $Q_i \neq P_i$ dividing p_i since $D \equiv 1 \bmod p$. We show how to compute integers $m_1, \ldots, m_k > 0$ with the property that m_e is the smallest value of $i_e > 0$ for which there exist $i_1, \ldots, i_{e-1}$ (which may be negative) such that $P_1^{i_1} \cdots P_e^{i_e}$ is principal. It is clear that the order of S is $m_1 \cdots m_k$.

First of all, notice that one can choose $i_1, \ldots, i_e$ so that $0 \leq i_j < m_j$, for $j < e$, and so that $P_1^{i_1} \cdots P_e^{i_e}$ is principal. In fact, suppose $0 \leq i_j < m_j$ for $e > j > s$. Write $i_s = \lambda m_s + \mu$, $0 \leq \mu < m_s$. There exists a principal ideal $I = P_1^{f_1} \cdots P_s^{m_s}$, and so $(I)^{-\lambda} P_1^{i_1} \cdots P_e^{i_e} = P_1^{i_1'} \cdots P_e^{i_e'}$ is principal. But $i_j' = i_j$ for $j > s$, while $i_s' = \mu$. Proceeding in this way, we obtain the desired form.

Let m_e' be the minimum value of $\lambda_e > 0$ for which there exist nonnegative $\lambda_1, \ldots, \lambda_{e-1}$ $(\lambda_i < m_i)$ such that the equation:

$$(*) \qquad x^2 + (4a - 1)y^2 = 4p_1^{\lambda_1} \cdots p_e^{\lambda_e}$$

has an integral solution with $x \equiv y \bmod 2$ and $y \neq 0$.

<u>Claim</u>: $m_e = m_e'$.

This will imply that m_e is effectively computable, since $4a - 1 > 0$ and so (*) has only to be checked for a finite number of x, y (assuming $m_1, \ldots, m_{e-1}$ are known).

Choose a solution of (*) with $\lambda_i \geq 0$ and $\lambda_e = m_e'$, and then set $\alpha = \frac{1}{2}(x + y\sqrt{1 - 4a})$ which is an integer. Now $N(\alpha) = \frac{1}{4}(x^2 + 4a - 1)y^2) = p_1^{\lambda_1} \cdots p_e^{\lambda_e}$. Thus the principal ideal (α)

must have prime decomposition $P_1^{\sigma_1}Q_1^{\tau_1}\dots P_e^{\sigma_e}Q_e^{\tau_e}$, where $\sigma_i, \tau_i \geq 0$ and $\sigma_i + \tau_i = \lambda_i$, as can be seen by looking at the norm. Since P_iQ_i is principal, it follows that $P_1^{\sigma_1-\tau_1}\dots P_e^{\sigma_e-\tau_e}$ is principal. By definition of m_e, we have $m_e \leq |\sigma_e - \tau_e| \leq \lambda_e = m_e'$.

On the other hand, suppose $P_1^{i_1}\dots P_e^{i_e}$ is principal, where $i_e = m_e$ and $i_j \geq 0$, as is shown above. If $(\alpha) = P_1^{i_1}\dots P_e^{i_e}$, write $\alpha = \frac{1}{2}(x + y\sqrt{1 - 4a})$, where $x \equiv y \bmod 2$. Note that $y \neq 0$, or else $Q_e^{m_e}$ would also have to occur in the prime factorization of (α). Now $\frac{1}{4}(x^2 + (4a - 1)y^2) = N(\alpha) = p_1^{i_1}\dots p_e^{i_e}$. By definition of m_e', we have $m_e' \leq i_e = m_e$ and the Claim is proved.

The above procedure is abetted by the following:

<u>PROPOSITION</u>: If $a = p_1^{e_i}\dots p_k^{e_k}$, then $P_1^{a_1}\dots P_k^{a_k}$ is <u>not</u> principal if $\Pi p_i^{|a_i|} < \Pi p_i^{e_i}$. Moreover, for some $a_i = \pm e_i$, $P_1^{a_1}\dots P_k^{a_k}$ is principal.

As a consequence $|S| \geq e_1\dots e_k$, since it follows easily from the proposition that every $P_1^{a_1}\dots P_k^{a_k}$, where $0 \leq a_i < e_i$, represents a different element of S. If $k = 1$, it follows that S is cyclic of order e_1.

To prove the Proposition, we merely note that $P_1^{a_1}\dots P_k^{a_k}$ principal implies that the equation $x^2 + (4a - 1)y^2 = 4p_1^{|a_1|}\dots p_k^{|a_k|}$ has a solution with $y \neq 0$. This cannot be true if $4p_1^{|a_1|}\dots p_k^{|a_k|} < 4a$. Finally, note that $\frac{1}{2}(1 + \sqrt{1 - 4a})$ has norm a.

The results of computing the cases $a \leq 125$, which correspond to the tabulations of [B], is that the ideal class group of R_a is trivial except for the following:

order 2 when a = 13, 23, 29, 31, 47, 49, 58, 59, 64, 67, 100, 101, 121

order 3 when a = 53, 71, 83

order 4 when a = 73, 89, 109

order 6 when a = 103, 113

REFERENCES

[B] Z. Borevich, I. Shafarevich: Number Theory. New York: Academic Press, 1966.

[C] R. Crowell: The Group G'/G" of a Knot Group G, Duke Math. J. 30 (1963), 349-54.

[CR] C. Curtis, I. Reiner: Representation Theory of Finite Groups and Associative Algebras. New York: Wiley-Interscience, 1962.

[G] M. Gutierrez: On Knot Modules, Invent. Math. 17 (1972), 329-35.

[Hi] F. Hirzebruch, W. Neumann, S. Koh: Differentiable Manifolds and Quadratic Forms. New York: Dekker, 1971.

[J] R. Jacobowitz: Hermitian Forms over Global Fields, Amer. J. Math. 84 (1962), 441-65.

[L] J. Levine: Knot Modules, I, Trans. Amer. Math. Soc., 229 (1977), 1-50.

[L1] J. Levine: Polynomial Invariants of Knots of Codimension Two, Annals of Math. 84 (1966), 537-54.

[La] W. Landherr: Aquivilenz Hermitescher Formen Uber Einem Beliebigen Algebraischen Zahlkorper, Abh. Math. Sem. Hamburg Univ. 11 (1935), 245-8.

[MI] J. Milnor: On Isometries of Inner Product Spaces, Invent. Math. 8 (1969), 83-97.

[O] O. O'Meara: Introduction to Quadratic Forms. New York: Academic Press, 1963.

[S] J. Serre: Corps Locaux. Paris: Hermann, 1968.

[V] B. Van der Waerden: Algebra. New York: Ungar, 1970.

INDEX

Vol. 609: General Topology and Its Relations to Modern Analysis and Algebra IV. Proceedings 1976. Edited by J. Novák. XVIII, 225 pages. 1977.

Vol. 610: G. Jensen, Higher Order Contact of Submanifolds of Homogeneous Spaces. XII, 154 pages. 1977.

Vol. 611: M. Makkai and G. E. Reyes, First Order Categorical Logic. VIII, 301 pages. 1977.

Vol. 612: E. M. Kleinberg, Infinitary Combinatorics and the Axiom of Determinateness. VIII, 150 pages. 1977.

Vol. 613: E. Behrends et al., L^P-Structure in Real Banach Spaces. X, 108 pages. 1977.

Vol. 614: H. Yanagihara, Theory of Hopf Algebras Attached to Group Schemes. VIII, 308 pages. 1977.

Vol. 615: Turbulence Seminar, Proceedings 1976/77. Edited by P. Bernard and T. Ratiu. VI, 155 pages. 1977.

Vol. 616: Abelian Group Theory, 2nd New Mexico State University Conference, 1976. Proceedings. Edited by D. Arnold, R. Hunter and E. Walker. X, 423 pages. 1977.

Vol. 617: K. J. Devlin, The Axiom of Constructibility: A Guide for the Mathematician. VIII, 96 pages. 1977.

Vol. 618: I. I. Hirschman, Jr. and D. E. Hughes, Extreme Eigen Values of Toeplitz Operators. VI, 145 pages. 1977.

Vol. 619: Set Theory and Hierarchy Theory V, Bierutowice 1976. Edited by A. Lachlan, M. Srebrny, and A. Zarach. VIII, 358 pages. 1977.

Vol. 620: H. Popp, Moduli Theory and Classification Theory of Algebraic Varieties. VIII, 189 pages. 1977.

Vol. 621: Kauffman et al., The Deficiency Index Problem. VI, 112 pages. 1977.

Vol. 622: Combinatorial Mathematics V, Melbourne 1976. Proceedings. Edited by C. Little. VIII, 213 pages. 1977.

Vol. 623: I. Erdelyi and R. Lange, Spectral Decompositions on Banach Spaces. VIII, 122 pages. 1977.

Vol. 624: Y. Guivarc'h et al., Marches Aléatoires sur les Groupes de Lie. VIII, 292 pages. 1977.

Vol. 625: J. P. Alexander et al., Odd Order Group Actions and Witt Classification of Innerproducts. IV, 202 pages. 1977.

Vol. 626: Number Theory Day, New York 1976. Proceedings. Edited by M. B. Nathanson. VI, 241 pages. 1977.

Vol. 627: Modular Functions of One Variable VI, Bonn 1976. Proceedings. Edited by J.-P. Serre and D. B. Zagier. VI, 339 pages. 1977.

Vol. 628: H. J. Baues, Obstruction Theory on the Homotopy Classification of Maps. XII, 387 pages. 1977.

Vol. 629: W. A. Coppel, Dichotomies in Stability Theory. VI, 98 pages. 1978.

Vol. 630: Numerical Analysis, Proceedings, Biennial Conference, Dundee 1977. Edited by G. A. Watson. XII, 199 pages. 1978.

Vol. 631: Numerical Treatment of Differential Equations. Proceedings 1976. Edited by R. Bulirsch, R. D. Grigorieff, and J. Schröder. X, 219 pages. 1978.

Vol. 632: J.-F. Boutot, Schéma de Picard Local. X, 165 pages. 1978.

Vol. 633: N. R. Coleff and M. E. Herrera, Les Courants Résiduels Associés à une Forme Méromorphe. X, 211 pages. 1978.

Vol. 634: H. Kurke et al., Die Approximationseigenschaft lokaler Ringe. IV, 204 Seiten. 1978.

Vol. 635: T. Y. Lam, Serre's Conjecture. XVI, 227 pages. 1978.

Vol. 636: Journées de Statistique des Processus Stochastiques, Grenoble 1977, Proceedings. Edité par Didier Dacunha-Castelle et Bernard Van Cutsem. VII, 202 pages. 1978.

Vol. 637: W. B. Jurkat, Meromorphe Differentialgleichungen. VII, 194 Seiten. 1978.

Vol. 638: P. Shanahan, The Atiyah-Singer Index Theorem, An Introduction. V, 224 pages. 1978.

Vol. 639: N. Adasch et al., Topological Vector Spaces. V, 125 pages. 1978.

Vol. 640: J. L. Dupont, Curvature and Characteristic Classes. X, 175 pages. 1978.

Vol. 641: Séminaire d'Algèbre Paul Dubreil, Proceedings Paris 1976–1977. Edité par M. P. Malliavin. IV, 367 pages. 1978.

Vol. 642: Theory and Applications of Graphs, Proceedings, Michigan 1976. Edited by Y. Alavi and D. R. Lick. XIV, 635 pages. 1978.

Vol. 643: M. Davis, Multiaxial Actions on Manifolds. VI, 141 pages. 1978.

Vol. 644: Vector Space Measures and Applications I, Proceedings 1977. Edited by R. M. Aron and S. Dineen. VIII, 451 pages. 1978.

Vol. 645: Vector Space Measures and Applications II, Proceedings 1977. Edited by R. M. Aron and S. Dineen. VIII, 218 pages. 1978.

Vol. 646: O. Tammi, Extremum Problems for Bounded Univalent Functions. VIII, 313 pages. 1978.

Vol. 647: L. J. Ratliff, Jr., Chain Conjectures in Ring Theory. VIII, 133 pages. 1978.

Vol. 648: Nonlinear Partial Differential Equations and Applications, Proceedings, Indiana 1976–1977. Edited by J. M. Chadam. VI, 206 pages. 1978.

Vol. 649: Séminaire de Probabilités XII, Proceedings, Strasbourg, 1976–1977. Edité par C. Dellacherie, P. A. Meyer et M. Weil. VIII, 805 pages. 1978.

Vol. 650: C*-Algebras and Applications to Physics. Proceedings 1977. Edited by H. Araki and R. V. Kadison. V, 192 pages. 1978.

Vol. 651: P. W. Michor, Functors and Categories of Banach Spaces. VI, 99 pages. 1978.

Vol. 652: Differential Topology, Foliations and Gelfand-Fuks-Cohomology, Proceedings 1976. Edited by P. A. Schweitzer. XIV, 252 pages. 1978.

Vol. 653: Locally Interacting Systems and Their Application in Biology. Proceedings, 1976. Edited by R. L. Dobrushin, V. I. Kryukov and A. L. Toom. XI, 202 pages. 1978.

Vol. 654: J. P. Buhler, Icosahedral Golois Representations. III, 143 pages. 1978.

Vol. 655: R. Baeza, Quadratic Forms Over Semilocal Rings. VI, 199 pages. 1978.

Vol. 656: Probability Theory on Vector Spaces. Proceedings, 1977. Edited by A. Weron. VIII, 274 pages. 1978.

Vol. 657: Geometric Applications of Homotopy Theory I, Proceedings 1977. Edited by M. G. Barratt and M. E. Mahowald. VIII, 459 pages. 1978.

Vol. 658: Geometric Applications of Homotopy Theory II, Proceedings 1977. Edited by M. G. Barratt and M. E. Mahowald. VIII, 487 pages. 1978.

Vol. 659: Bruckner, Differentiation of Real Functions. X, 247 pages. 1978.

Vol. 660: Equations aux Dérivée Partielles. Proceedings, 1977. Edité par Pham The Lai. VI, 216 pages. 1978.

Vol. 661: P. T. Johnstone, R. Paré, R. D. Rosebrugh, D. Schumacher, R. J. Wood, and G. C. Wraith, Indexed Categories and Their Applications. VII, 260 pages. 1978.

Vol. 662: Akin, The Metric Theory of Banach Manifolds. XIX, 306 pages. 1978.

Vol. 663: J. F. Berglund, H. D. Junghenn, P. Milnes, Compact Right Topological Semigroups and Generalizations of Almost Periodicity. X, 243 pages. 1978.

Vol. 664: Algebraic and Geometric Topology, Proceedings, 1977. Edited by K. C. Millett. XI, 240 pages. 1978.

Vol. 665: Journées d'Analyse Non Linéaire. Proceedings, 1977. Edité par P. Bénilan et J. Robert. VIII, 256 pages. 1978.

Vol. 666: B. Beauzamy, Espaces d'Interpolation Réels: Topologie et Géometrie. X, 104 pages. 1978.

Vol. 667: J. Gilewicz, Approximants de Padé. XIV, 511 pages. 1978.

Vol. 668: The Structure of Attractors in Dynamical Systems. Proceedings, 1977. Edited by J. C. Martin, N. G. Markley and W. Perrizo. VI, 264 pages. 1978.

Vol. 669: Higher Set Theory. Proceedings, 1977. Edited by G. H. Müller and D. S. Scott. XII, 476 pages. 1978.

Vol. 670: Fonctions de Plusieurs Variables Complexes III, Proceedings, 1977. Edité par F. Norguet. XII, 394 pages. 1978.

Vol. 671: R. T. Smythe and J. C. Wierman, First-Passage Perculation on the Square Lattice. VIII, 196 pages. 1978.

Vol. 672: R. L. Taylor, Stochastic Convergence of Weighted Sums of Random Elements in Linear Spaces. VII, 216 pages. 1978.

Vol. 673: Algebraic Topology, Proceedings 1977. Edited by P. Hoffman, R. Piccinini and D. Sjerve. VI, 278 pages. 1978.

Vol. 674: Z. Fiedorowicz and S. Priddy, Homology of Classical Groups Over Finite Fields and Their Associated Infinite Loop Spaces. VI, 434 pages. 1978.

Vol. 675: J. Galambos and S. Kotz, Characterizations of Probability Distributions. VIII, 169 pages. 1978.

Vol. 676: Differential Geometrical Methods in Mathematical Physics II, Proceedings, 1977. Edited by K. Bleuler, H. R. Petry and A. Reetz. VI, 626 pages. 1978.

Vol. 677: Séminaire Bourbaki, vol. 1976/77, Exposés 489–506. IV, 264 pages. 1978.

Vol. 678: D. Dacunha-Castelle, H. Heyer et B. Roynette. Ecole d'Eté de Probabilités de Saint-Flour. VII-1977. Edité par P. L. Hennequin. IX, 379 pages. 1978.

Vol. 679: Numerical Treatment of Differential Equations in Applications, Proceedings, 1977. Edited by R. Ansorge and W. Törnig. IX, 163 pages. 1978.

Vol. 680: Mathematical Control Theory, Proceedings, 1977. Edited by W. A. Coppel. IX, 257 pages. 1978.

Vol. 681: Séminaire de Théorie du Potentiel Paris, No. 3, Directeurs: M. Brelot, G. Choquet et J. Deny. Rédacteurs: F. Hirsch et G. Mokobodzki. VII, 294 pages. 1978.

Vol. 682: G. D. James, The Representation Theory of the Symmetric Groups. V, 156 pages. 1978.

Vol. 683: Variétés Analytiques Compactes, Proceedings, 1977. Edité par Y. Hervier et A. Hirschowitz. V, 248 pages. 1978.

Vol. 684: E. E. Rosinger, Distributions and Nonlinear Partial Differential Equations. XI, 146 pages. 1978.

Vol. 685: Knot Theory, Proceedings, 1977. Edited by J. C. Hausmann. VII, 311 pages. 1978.

Vol. 686: Combinatorial Mathematics, Proceedings, 1977. Edited by D. A. Holton and J. Seberry. IX, 353 pages. 1978.

Vol. 687: Algebraic Geometry, Proceedings, 1977. Edited by L. D. Olson. V, 244 pages. 1978.

Vol. 688: J. Dydak and J. Segal, Shape Theory. VI, 150 pages. 1978.

Vol. 689: Cabal Seminar 76–77, Proceedings, 1976–77. Edited by A.S. Kechris and Y. N. Moschovakis. V, 282 pages. 1978.

Vol. 690: W. J. J. Rey, Robust Statistical Methods. VI, 128 pages. 1978.

Vol. 691: G. Viennot, Algèbres de Lie Libres et Monoïdes Libres. III, 124 pages. 1978.

Vol. 692: T. Husain and S. M. Khaleelulla, Barrelledness in Topological and Ordered Vector Spaces. IX, 258 pages. 1978.

Vol. 693: Hilbert Space Operators, Proceedings, 1977. Edited by J. M. Bachar Jr. and D. W. Hadwin. VIII, 184 pages. 1978.

Vol. 694: Séminaire Pierre Lelong – Henri Skoda (Analyse) Année 1976/77. VII, 334 pages. 1978.

Vol. 695: Measure Theory Applications to Stochastic Analysis, Proceedings, 1977. Edited by G. Kallianpur and D. Kölzow. XII, 261 pages. 1978.

Vol. 696: P. J. Feinsilver, Special Functions, Probability Semigroups, and Hamiltonian Flows. VI, 112 pages. 1978.

Vol. 697: Topics in Algebra, Proceedings, 1978. Edited by M. F. Newman. XI, 229 pages. 1978.

Vol. 698: E. Grosswald, Bessel Polynomials. XIV, 182 pages. 1978.

Vol. 699: R. E. Greene and H.-H. Wu, Function Theory on Manifolds Which Possess a Pole. III, 215 pages. 1979.

Vol. 700: Module Theory, Proceedings, 1977. Edited by C. Faith and S. Wiegand. X, 239 pages. 1979.

Vol. 701: Functional Analysis Methods in Numerical Analysis, Proceedings, 1977. Edited by M. Zuhair Nashed. VII, 333 pages. 1979.

Vol. 702: Yuri N. Bibikov, Local Theory of Nonlinear Analytic Ordinary Differential Equations. IX, 147 pages. 1979.

Vol. 703: Equadiff IV, Proceedings, 1977. Edited by J. Fábera. XIX, 441 pages. 1979.

Vol. 704: Computing Methods in Applied Sciences and Engineering, 1977, I. Proceedings, 1977. Edited by R. Glowinski and J. L. Lions. VI, 391 pages. 1979.

Vol. 705: O. Forster und K. Knorr, Konstruktion verseller Familien kompakter komplexer Räume. VII, 141 Seiten. 1979.

Vol. 706: Probability Measures on Groups, Proceedings, 1978. Edited by H. Heyer. XIII, 348 pages. 1979.

Vol. 707: R. Zielke, Discontinuous Čebyšev Systems. VI, 111 pages. 1979.

Vol. 708: J. P. Jouanolou, Equations de Pfaff algébriques. V, 255 pages. 1979.

Vol. 709: Probability in Banach Spaces II. Proceedings, 1978. Edited by A. Beck. V, 205 pages. 1979.

Vol. 710: Séminaire Bourbaki vol. 1977/78, Exposés 507–524. IV, 328 pages. 1979.

Vol. 711: Asymptotic Analysis. Edited by F. Verhulst. V, 240 pages. 1979.

Vol. 712: Equations Différentielles et Systèmes de Pfaff dans le Champ Complexe. Edité par R. Gérard et J.-P. Ramis. V, 364 pages. 1979.

Vol. 713: Séminaire de Théorie du Potentiel, Paris No. 4. Edité par F. Hirsch et G. Mokobodzki. VII, 281 pages. 1979.

Vol. 714: J. Jacod, Calcul Stochastique et Problèmes de Martingales. X, 539 pages. 1979.

Vol. 715: Inder Bir S. Passi, Group Rings and Their Augmentation Ideals. VI, 137 pages. 1979.

Vol. 716: M. A. Scheunert, The Theory of Lie Superalgebras. X, 271 pages. 1979.

Vol. 717: Grosser, Bidualräume und Vervollständigungen von Banachmoduln. III, 209 pages. 1979.

Vol. 718: J. Ferrante and C. W. Rackoff, The Computational Complexity of Logical Theories. X, 243 pages. 1979.

Vol. 719: Categorial Topology, Proceedings, 1978. Edited by H. Herrlich and G. Preuß. XII, 420 pages. 1979.

Vol. 720: E. Dubinsky, The Structure of Nuclear Fréchet Spaces. V, 187 pages. 1979.

Vol. 721: Séminaire de Probabilités XIII. Proceedings, Strasbourg, 1977/78. Edité par C. Dellacherie, P. A. Meyer et M. Weil. VII, 647 pages. 1979.

Vol. 722: Topology of Low-Dimensional Manifolds. Proceedings, 1977. Edited by R. Fenn. VI, 154 pages. 1979.

Vol. 723: W. Brandal, Commutative Rings whose Finitely Generated Modules Decompose. II, 116 pages. 1979.

Vol. 724: D. Griffeath, Additive and Cancellative Interacting Particle Systems. V, 108 pages. 1979.

Vol. 725: Algèbres d'Opérateurs. Proceedings, 1978. Edité par P. de la Harpe. VII, 309 pages. 1979.

Vol. 726: Y.-C. Wong, Schwartz Spaces, Nuclear Spaces and Tensor Products. VI, 418 pages. 1979.

Vol. 727: Y. Saito, Spectral Representations for Schrödinger Operators With Long-Range Potentials. V, 149 pages. 1979.

Vol. 728: Non-Commutative Harmonic Analysis. Proceedings, 1978. Edited by J. Carmona and M. Vergne. V, 244 pages. 1979.